HOW TO PREVENT DEMENTIA

UNDERSTANDING AND MANAGING COGNITIVE DECLINE

RICHARD RESTAK, MD

Skyhorse Publishing

Skyhorse Publishing books may be purchased in bulk at special discounts for sales promotion, corporate gifts, fund-raising, or educational purposes. Special editions can also be created to specifications. For details, contact the Special Sales Department, Skyhorse Publishing, 307 West 36th Street, 11th Floor, New York, NY 10018 or info@skyhorsepublishing.com.

Skyhorse® and Skyhorse Publishing® are registered trademarks of Skyhorse Publishing, Inc.®, a Delaware corporation.

Visit our website at www.skyhorsepublishing.com
Please follow our publisher Tony Lyons on Instagram @tonylyonsisuncertain

10 9 8 7 6 5 4 3 2 1

Library of Congress Cataloging-in-Publication Data is available on file.

Hardcover ISBN: 978-1-5107-7629-6
Ebook ISBN: 978-1-5107-7630-2

Cover design by Brian Peterson

Printed in the United States of America

To Carolyn, Jennifer, Alison, and Ann—
the four most important women in my life.

CONTENTS

CHAPTER I

INTRODUCTION TO DEMENTIA

Several core beliefs underlie the structure of this book. First, the more you know about Alzheimer's disease and the other dementias, the more tools you will have at your disposal to prevent its onset in yourself and be of assistance to others who are less fortunate. If an illness isn't understood, it's only too easy to become anxious when encountering or even hearing about it. I'm not suggesting you have to achieve a neuropsychiatrist's level of understanding of the dementias, but I'm convinced that the more you know about them, the better.

The second belief emphasizes the importance of considering the thinking disturbances associated with the dementias. This is important when it comes to how we think about the origin and treatment of the disease. It's commonly believed that Alzheimer's primarily affects memory, and in most cases that's correct. But Alzheimer's can also begin with speech problems, both understanding what other people say and speaking in a manner that others can understand. Alzheimer's and the other dementias may involve disorders of emotions and behavior: unreasonable anxieties and depressions, reclusiveness, hording, impatience, sudden flairs of temper, delusions, and hallucinations. But even these extreme behaviors are the expressions of *disorders of thought*.

The final belief concerns the existence of a continuum from normal thought to dementia. When speaking or writing about thought, psychologists employ the term *cognition*, which includes orientation, attention, language, abstraction, memory, naming, and visualization. If we consider Alzheimer's and other dementias as thinking disorders, we find that the illnesses are not mysterious maladies unrelated to ordinary life. On occasion we all experience disorders in our thinking: we find it difficult to concentrate, to visualize, to come up with names. We get lost in neighborhoods we are not familiar with. On occasion we can become mildly suspicious of other people's motives and, rarely, see movements or figures appearing fleetingly in our peripheral vision when we know there is nothing there. By considering Alzheimer's and the other dementias as existing on a continuum extending from normal thinking to severe mental impairment, the illness becomes easier to understand and less anxiety provoking ("Am I coming down with the big A?").

Despite the fact that dementia was first recognized as a brain disease by Alzheimer almost a century and a quarter ago, we really cannot answer the basic question "What is the cause of Alzheimer's?" Although this seems like a purely scientific question, it isn't. As I will discuss in specific sections of this book, progress, understanding, and treatment for the illness is also impeded by social, cultural, economic, and statistical issues.

A BIRD'S-EYE VIEW

Before we get to ways of preventing or diminishing dementia, let's first define it and distinguish it from Alzheimer's. The two terms are commonly confused.

Think of dementia as an umbrella term that contains many different diseases, just as the term animal refers to a myriad number of creatures. All cats and dogs are animals. But not all animals are cats or dogs.

Dementia represents a decline in mental function that can result from several causes, some curable, some not. At the moment,

Alzheimer's is the most commonly encountered incurable form of dementia. While I personally believe a remedy will developed in the next five to ten years, what can we do in the meantime? That is the topic of this book.

Let's start with a bird's-eye view of Alzheimer Land. Here are the naked statistics about its prevalence and the risk of coming down with it.

Alzheimer's statistics, like statistical approaches to almost anything, provide justification for the pessimist and hope for the optimist. An estimated 6 to 6.7 million Americans age 65 and older are living with Alzheimer's in 2023, with 73 percent of them age 75 or older. The incidence in 2025 (almost tomorrow) is expected to reach 7.1 million people, a 27 percent increase from 5.6 million age 65 and older in 2019. If you extrapolate the numbers even further into the future, the percentage becomes even more panic-inducing.

By 2050, or earlier in the absence of a breakthrough, preventions, or cures, the number of people age 65 and older with Alzheimer's is projected to reach 12.7 million people. Deaths from Alzheimer's have more than doubled between 2000 and 2019, while those from heart disease—the leading cause of death—have decreased. Alzheimer deaths between 2000 and 2019 increased by 145 percent. Alzheimer's or other dementias kill more than breast cancer and prostate cancer combined.

Of course, all these figures have to be taken with a good deal of reservation, since currently only one in four people with Alzheimer's are diagnosed. Despite this low recognition factor, Alzheimer's remains (depending on whom you ask) the sixth or seventh leading cause of death in the United States. What's even more sobering, Alzheimer's is the only disease among the ten leading causes of death that currently cannot be cured, totally prevented, or reliably slowed in its progression.

But I don't see any value in continuing such doom-saying since a lot of current research looks promising. Further, as we progress in this book, I will mention the steps that can be taken *now* to reduce the risk of Alzheimer's and the other dementias.

Inevitably the question arises: "Is Alzheimer's more prevalent now than in the past, even the recent past?" Before jumping to any conclusions, consider this. In the first quarter of the twentieth century, the period when Alzheimer's was first diagnosed, the average life expectancy was 47.3 years. Put another way, the life span has about doubled in the past century. As a result, the disease was formerly rare except for the early-onset form, where the person with Alzheimer's carries a gene or genes leading to the illness. Indeed, the world's first identified case was a woman who came down with the disease at 51 years old and died at 57. By present criteria, she probably had early-onset Alzheimer's (before age 65). So the question about the frequency of Alzheimer's disease in earlier times is unanswerable. Only in a culture with an average lifespan of 65 or more years will Alzheimer's occupy a prominent position on a chart listing the most frequent causes of death.

The most common form of dementia worldwide, *late-onset Alzheimer's*, depends on the simple process of aging. The older you become, the greater the likelihood of coming down with the disease. Genetics may play some role here, but in general it is a less critical component than it is in the *early-onset* form of the illness.

Comprising only about 1 to 2 percent of Alzheimer's diagnoses, the early-onset form is predominantly inherited and typically occurs before age 65. The children of Alzheimer's patients who have inherited the gene (or genes) responsible for the illness (a 50 percent chance of inheritance) are highly likely to come down with Alzheimer's during their productive years (ages 20-60).

Putting a more positive spin on it, if your family history is "clean," you have a very slight prospect of contracting Alzheimer's before age 65. Based on this fact alone, you shouldn't be needlessly fretting about Alzheimer's in your thirties or forties as many people currently are.

In early 2020, the World Health Organization released a report, Global Health Estimates, listing the top ten causes of death from 2000 to 2019. Although Alzheimer's ranked as the seventh leading cause of death globally, it didn't occur homogeneously across global income groups.

The more advanced and economically secure a country, the greater the risk of their citizens succumbing to dementias. (Since COVID-19 exerted such a powerful effect on the death rate within all income groups from 2020 onward, the figures I'm quoting are from the World Health Organization's Top 10 Causes of Death from 2000 to 2019.) What presently unknown factors in upper-middle-income countries are contributing to the meteoric rise in incidence in the ten top causes, ranging from not even listed to eighth among upper-income countries?

Longevity is one obvious factor, resulting from better diet, the availability (at least among certain sections of the population) of excellent health care, cleaner air and water, etc. If you live in a high-income country, the average lifespan is longer (discounting for the moment such contributing factors as suicide, drug overdoses, and violence). So, yes, longevity is an important factor in increasing the incidence of Alzheimer's and other dementias within our society. But longevity isn't a totally satisfying explanation. An unknown number of contributing factors, plus our ignorance of the cause of the disease, will likely make a cure elusive.

THE FOUR A'S

Four impairments underlie the outer expressions and inner experiences of the Alzheimer patient. Since the word for each impairment begins with the letter A, let's describe them together as the **four A's.**

Amnesia refers to memory loss, typically beginning with short-term memory.

Aphasia involves difficulty finding the right words or using those words incorrectly (expressive aphasia) or failing to understand or interpret language spoken by somebody else (receptive aphasia).

Obviously, neither amnesia nor aphasia in their milder forms is always abnormal. Who doesn't occasionally forget? Or experience difficulty coming up with the right word, or totally comprehending the words that someone else is saying? So, the first two of the four A's fit quite comfortably on a continuum model, ranging from normal mild amnesia and aphasia (occasionally present in anyone, especially under circumstances of fatigue or stress) and extending on the other end of the continuum to someone visibly and seriously impaired.

But the third and fourth of the four A's are always indicators that something is amiss.

Apraxia, taken from the Greek words *a* (without) and *praxis* (action), is the term for difficulty or inability performing purposeful and highly practiced actions despite normal muscle strength and tone. A person with apraxia may be able to recognize and even name a toothbrush and toothpaste but may be unable to carry out the act (praxis) of squeezing the toothpaste onto the toothbrush. Or the apraxia may involve an inability to put the toothbrush in the mouth and perform the movements necessary to brush the teeth. All of the muscle components are present but can't be coordinated. The apraxic person can be unkindly described as a person who literally can't get his act together.

If the apraxia involves the legs and arms, the person may sway and fall to the ground—one contributor to the increased incidence of falls and fractures, especially of the hip. In the later stages of Alzheimer's, apraxia explains why the affected person can't make a meal, get dressed, or wash himself or herself.

Apraxia can also affect speech. Despite normal tongue and mouth movements and a desire to speak, a person with speech apraxia experiences great difficulty or even total inability to move their mouth and tongue to produce understandable speech. In addition to apraxia of speech, a form of apraxia may occur (orofacial) that robs the person of an ability to perform certain facial movements, such as winking or symmetrically moving both sides of the face.

Finally, **agnosia** is again from two Greek words: *a* (without) and *gnosis* (knowledge). Agnosia refers to an impairment in correctly understanding information provided by the senses of seeing, hearing, touching, smelling, and tasting. With agnosia of sight, a spouse or other family member may not be recognized by vision alone. Or a sensation such as hearing may be intact, but the meaning of what is heard may be impaired: the screech of breaks and a loud horn may be not recognized soon enough to avoid being struck by a fast-approaching car.

Many, if not all, expressions of Alzheimer's can be explained by reference to the four A's.

Since the cause of Alzheimer's remains unknown, it should come as no surprise to learn that in most cases an onset event or starting point cannot be identified. One thing we know for sure is that the disease process begins long before the first appearance of symptoms. The initial stage is marked by an uncertain starting point described as *mild cognitive impairment* (MCI). MCI may or may not be the initial starting point for Alzheimer's disease; only the passage of time can permit that determination.

Initially the symptoms of MCI are barely noticeable, except to the keen observer and may only involve a mild decline in thinking, occurring in a setting of overall generally acceptable function. The person with MCI can come and go to the supermarket, for instance, but must write down a grocery list; nor can the person remember, as done previously, the aisle in which a particular grocery item can be found. If only the memory is involved, it's called *amnestic MCI* and can be the initial stage of Alzheimer's or may not progress at all.

One of the patients of Robert Peterson, a neurologist at the Mayo Clinic, is a 70-year-old businessman with MCI. He can successfully handle computer operations, as well as manage his professional and personal finances. He still functions adequately on various committees and boards but has to take notes to remember details. Of particular concern is the development of increasing forgetfulness, especially when trying to come up with people's names and recent events. And although he remains reasonable and companionable, he has started experiencing and expressing mild irritability. On neuropsychological tests, he does well.

As with Peterson's patient, people with MCI often show only mild impairments in memory, language, and decision-making. MCI incidence increases with age. According to the American Academy of Neurology, MCI affects about 8 percent of people in the ages 65–69 range; 10 percent of those in the 70–74 range, and 15 percent of people 75–79 years of age. Over a third of people aged 85 and older are affected with MCI. Other *symptoms* (complaints from patients or others) and *signs* (observable objective expressions of decline from previous functioning) consist of difficulty judging the amount of time required to arrive promptly for appointments or misjudging the sequence of steps necessary to carry out a specific task, like arranging a car trip or a vacation.

Some of those affected with MCI may progress to more definite stages of Alzheimer's; others may not progress in the short term but over a few years may display undeniable signs of the disease. None of them returns to their prior status. Unfortunately, at this point the best that can be hoped for with MCI is a stabilizing of this so-far mysterious disease process.

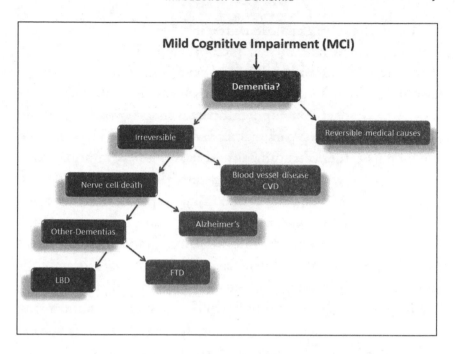

Alzheimer's Disease (AD)

AD is the most common neurodegenerative disorder and cause of dementia.

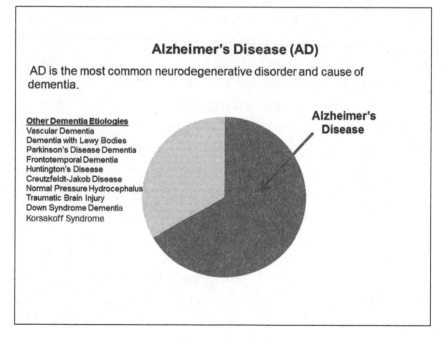

Other Dementia Etiologies
Vascular Dementia
Dementia with Lewy Bodies
Parkinson's Disease Dementia
Frontotemporal Dementia
Huntington's Disease
Creutzfeldt-Jakob Disease
Normal Pressure Hydrocephalus
Traumatic Brain Injury
Down Syndrome Dementia
Korsakoff Syndrome

Alzheimer's Disease

If MCI progresses to Alzheimer's, it does so in three stages:

Mild dementia: Efficient memory and thinking are usually first affected and involve additional amnesia and aphasia, the first two of the four A's. Although noticeable, these mild failures do not severely affect home or work relationships. Most notable are failures to remember conversations that were held days or even only hours earlier. This difficulty in learning new information is often coupled with repetitive questioning about the information that has already been provided. As a consequence of apraxic difficulties, family events prove overly burdensome and stressful. Balancing a checkbook becomes well-nigh impossible as a result of the increasing apraxia. Financial decisions are put off or improperly managed—resulting in lost money and conflicts with family members regarding faulty financial decisions. Accompanying such mismanagement are impulsive purchases, often involving items offered for sale on television shopping channels.

Other aphasic and apraxic difficulties include additional problems organizing and expressing thoughts, thanks to the loss of the correct words set out in the proper sequence; misplacing belongings in the home; easily losing one's way outside of the home; and loss of the usual "get up and go," leading to reduced motivation to carry out formerly pleasurable pursuits (apathy). Although many of these developments can be deeply troubling, the onset of Alzheimer's varies greatly from person to person. The transition to the moderate and severe forms may take anywhere from months to years.

Moderate dementia involves all four of the four A's. Increasing confusion and poor judgment are frequently coupled with wandering from the home. Greater memory loss leads to the inability to find even frequently used items within the home. At this stage, accusations are frequently made that people are stealing even disposable, inexpensive items. If not stealing, then "they" are accused of hiding the lost objects. Such delusional beliefs make it difficult to hire and retain workers to supervise and oversee problematic behaviors.

Severe dementia involves a nearly total loss of the ability to communicate coherently. Imagine a whiteboard in a classroom at the end of a history class. Before leaving the room, the teacher made a desultory and partly successful attempt to erase the board. What's left behind were half-erased, smudged words and phrases. Some words are no longer even recognizable as words but only as strings of letters loosely connected to one another. Sentences that had been perfectly readable and understandable only a few moments before the partial erasure now make no sense at all. Following this metaphor, advanced Alzheimer's disease is the result of the erasure randomly applied to the writing, resulting in a mélange of poorly shaped words and letters that fail to convey meaning.

This profound loss results in the need for twenty-four-hour assistance with eating and personal care; a severe decline in physical ability, leading to a bedridden state; culminating in death by pneumonia—secondary to a swallowing impairment, whereby food and drink enter the airways and the lungs with subsequent infection, sepsis (blood-born infection), and death. The time from first diagnosis to death varies greatly. People with Alzheimer's disease live anywhere from three years to about eleven years, with some few surviving twenty years or more.

The patterns and speed of onset of Alzheimer's depend on vulnerability and resiliency.

Let's talk about vulnerability first. Risk factors are divided into unmodifiable and modifiable. Fortunately, the modifiable risk factors outnumber the unmodifiable ones.

Unmodifiable risk factors are age, sex, and family history. None of these can be changed.

Modifiable risk factors include body mass, diabetes, sleep disorders, hypertension (high blood pressure), elevated cholesterol, depression, traumatic brain injury (TBI), smoking, drinking, and education. Improving any or all of these contributors cuts down on the likelihood of developing Alzheimer's. Ending the smoking habit alone makes someone 60 percent less likely to come down with Alzheimer's.

By combining these risk factors, we come up with a composite of the person least likely to come down with Alzheimer's: a non-smoker, nondrinker, and physically active person, free of diabetes and depression, with normal body mass, cholesterol, and mental curiosity. I'll largely leave to your internist to manage the diabetes, body mass, hypertension, cholesterol, smoking, and drinking. I will concentrate throughout the remainder of the book on diet, exercise, sleep, depression, traumatic brain injury, education, and the other factors that exert a powerful influence on the likelihood of Alzheimer's.

TWO SETS OF QUESTIONS YOU CAN ASK TO DETECT EARLY DEMENTIA

Once we reach adulthood, our behavior has taken on a regularity and, in most cases, some degree of predictability (just ask your spouse if you want more information about that). Alzheimer's and other dementias represent perturbations in that regularity and predictability. In fact, the initial diagnosis almost never takes place in the absence of some notable failures of memory combined with behavioral changes. Here are two sets of questions anyone can use to detect the likely onset of Alzheimer's or other dementia in another person:

1. Does the person show a change in usual demeanor? Is the person displaying a recent onset of memory difficulty? Do they show a notable change in usual behavior? Dramatic or out-of-character behavior is always suspect because, as mentioned, most people by adulthood have established a distinct behavioral repertoire.

2. If the answer to any of the prior questions is yes, does that change in behavior interfere with daily living, especially relationships with others? Techniques for managing inner personal stress and conflict are established fairly early in life and maintained over our lifetime. That isn't to say that all of us can't fly off the handle now and again. But repetitive and emotionally excessive expressions (shouting or even more aggressive behavior, including physical outbursts) rarely increase in frequency during normal aging. They are, however, frequent and serious in the dementias.

If the answer is yes to both questions 1 and 2, neuropsychiatric illness—either dementia or psychiatric disease (most likely severe depression)—is highly probable.

If the answer to question 2 is no, the person most likely has mild cognitive impairment (MCI) or is normal (but only if the answers to 1 are also no).

The use of these two questions can be extremely helpful in family discussions about a relative, especially when not everyone is convinced there is a problem.

Once dementia is determined to be likely, one may be dealing with *reversible* forms of dementia, which result from medical causes, toxins, or poisons. Alcohol and drug intoxication is included here as well. Prompt diagnosis and treatment are paramount since the progression to dementia can be halted. Adjust the blood sugar or the kidney function, and remove the toxin or the poison, and the affected person recovers from the temporary dementia (termed *delirium*).

While thinking, understanding, and judgment can be affected in both dementia and delirium, delirium is distinguished by the rapidity of its onset, sometimes within a few hours. It also passes (often equally rapidly) after treatment for the cause.

Unfortunately, most cases of dementia are *irreversible* and involve either brain-cell death or a disease of the arteries carrying blood to the brain (*vascular cognitive impairment*).

Dementia secondary to nerve-cell death represents either Alzheimer's—at least two-thirds of the dementias—or *non-Alzheimer's*, which is broken down into separate discrete dementias, which will be taken up later. You may be wondering, *How do we know all this? When and how did neuroscientists learn how to recognize Alzheimer's?*

The answers to these important questions are best addressed by a short summary of the history of Alzheimer's disease.

CHAPTER II

EVOLUTION OF OUR UNDERSTANDING OF ALZHEIMER'S

NIGHT TRAIN TO BRESLAU

In early December 1915, a night train from Bonn, Germany, to Breslau carried a passenger whose name would be known worldwide by the end of the century. Alois Alzheimer, a portly 51-year-old psychiatrist, was traveling back to Breslau, where he had been professor of psychiatry since 1912. Although he often appeared forbidding, thanks to a penchant for heavily starched shirts, formal dress, and pince-nez glasses, he also loved impersonations and practical jokes. During his early training in the hospital, he disguised himself for Christmas parties as a peddler carrying a tray of toys to entertain his younger patients.

If Alzheimer tended to reminisce to the rhythms of the train speeding along the 436-mile journey to Breslau, he had much to reminisce about. While only in his early thirties, Alzheimer had already established himself as a thought leader—indeed *the* thought leader on the nature of a brain disease known as dementia. *Demens*, the etymological root for the word *dementia*, is a Latin adjective that means "out of one's mind," or "mad, raving, or insane."

As Alzheimer sat drowsily immersing himself in the clickety-click of the wheels, he gradually entered a twilight state between waking and sleeping. Against the background of this hypnogogic state, he thought of dementia and how over the previous two thousand years, millions of people considered it in terms of the humoral theory.

THE FOUR HUMORS

According to the humoral theory, chemical systems within the body regulate all aspects of human behavior. Although this concept originated with the Egyptians, it was not codified into a system until the Greeks, notably the philosopher and physician Hippocrates and Galen, the physician to the gladiators at Pergamum. Hippocrates believed that the predominance of one of four humors (blood, black bile, yellow bile, and phlegm) determined the personality of a particular individual. A preponderance of blood produced a sanguine (hopeful) temperament; yellow bile in overabundance led to an irritable temperament; black bile led to melancholy. The fourth humor, phlegm, was associated with the brain, based on the brain's color and consistency.

One text, *On the Constitution of the Universe and of Man*, speaks of the relationship between the physical elements of the universe (air, water, earth, fire) and the humoral elements making up the essence of humans. On the importance of phlegm and its relationship to dementia, the unknown author wrote, "Those who have phlegm are low spirited, forgetful and have white hair." Phlegm was a reasonable choice for explaining forgetfulness since it was also associated with winter, old age, and, when present in excess, senility.

The humoral theory began losing its appeal in the mid-nineteenth century with the advent of germ theory. In 1841 Ignaz Semmelweis, a Viennese obstetrician, noticed a difference in maternal deaths secondary to fevers following childbirth in women delivered by doctors compared to those attended by midwives. This discrepancy was just the opposite from what one would expect, considering the longer and more intense training of the obstetricians. The women delivered by midwives experienced far fewer deaths.

In an inspired insight, Semmelweis noted that the doctors often arrived in the delivery rooms after performing or assisting at an autopsy. Semmelweis asserted—to the vocal displeasure of many of his colleagues—that the post-childbirth fever resulted from contamination brought from the morgue to the operating room.

To test his theory, Semmelweis formulated a new rule in the delivery room. Doctors were required to wash their hands in chlorinated lime water before delivering or even examining pregnant women. During the next year, the death rate fell from 18 percent to 2 percent.

Over the next half of the nineteenth century, the germ theory achieved increasing acceptance. According to this theory, tiny invisible particles known as germs are responsible for infectious diseases. *Germ* can refer to a bacterium, a fungus, a parasite, or a virus. Thanks to this shift from humoral theory to germ theory, physicians realized that illnesses were associated with distinct causes. This didn't always imply an infectious origin; the majority of diseases aren't infectious. Generally, senility (an early term for dementia) was not thought to be infectious in origin. In regard to the brain, senility was no longer associated with a humor, but something affecting the brain directly. This raised the tantalizing question: what was the nature of this brain alteration leading to dementia?

Prior to the nineteenth century, dementia was based on a mix of folklore, superstition, and judgmentalism. Such descriptors as *idiocy*, *senility*, and *stupidity* prevailed. Overall, dementia was considered a form of madness.

Its regressive nature—culminating in a kind of second childhood—was depicted in Shakespeare's *King Lear* where Goneril laments what her father has become:

Idle old man
That still would manage those authorities
That he hath given away! Now, by my life,
Old fools are babes again.
(Act I, Scene II)

During the Renaissance, dementia was considered to be caused by a cooling of the brain. This arbitrary and false assumption remained popular into the later seventeenth century. But whatever the cause, most authorities persisted in defining dementia in terms usually applied to insanity.

Consistent with this emphasis on dementia as a species of madness, the two most famous nineteenth-century authorities (Alois Alzheimer and Emil Kraepelin) were both psychiatrists. Kraepelin is famous for recognizing and naming schizophrenia, which he called *Dementia praecox*, meaning "premature dementia" or "precocious madness," usually developing in the late teens or early adulthood. *Dementia praecox* (later referred to almost exclusively as schizophrenia) was marked by disturbances in thought and mental function, including attention, memory, and behavior.

In contrast to *Dementia praecox*, dementia—as we now think of it—involves an older person, usually with a lifetime of previously normal mental functioning. According to traditional thinking—not always correct—dementia almost exclusively affects the old, while schizophrenia is primarily a young person's disease.

THE INSANITY OF AUGUSTE DETER

In 1901, while working at a psychiatric hospital in Frankfurt, Germany, Alzheimer encountered a patient who would become his obsession for the rest of his life. Auguste Deter, age 51, was brought to the city Hospital for the Mentally Ill and Epileptics by her husband, Karl. Karl told Dr. Alzheimer that he was unable to keep a job anymore due to the disturbances caused by his wife. She would wake up in the middle of the night and scream for hours. Alzheimer's notes, available today, described a woman with a helpless expression. When shown various objects, she couldn't remember the names of them. Everything had to be repeated slowly and precisely for her to remember anything. "I've lost myself, so to say," was Auguste Deter's daily refrain.

Alzheimer was unable to identify Deter's illness. It seemed as if she was severely demented, but she was much too young for the dementia that accompanies extreme old age. Increasing paranoia culminated in the intractable belief that someone was trying to kill her. While in the hospital, Auguste continued to deteriorate with increasing confusion, disorientation, and delirium. In April 1906 Auguste Deter died at age 55.

Dr. Alzheimer had Auguste's brain brought to Munich, where he was working, and examined it under a microscope with the aid of a recently developed staining technique. Along with a generalized shrinkage of the brain, Alzheimer found two abnormal accumulations of protein, one within the cell (called *tangles*) and another in the spaces between the cells, called *plaques*. As Alzheimer and others were to observe, these two abnormal protein aggregates were present in cases with conditions similar to Auguste Deter's.

In opposition to Alzheimer's view of mental illness was the towering figure of Sigmund Freud. Although Freud was originally a neurologist, he abruptly underwent a change in his interests and beliefs around the turn of the century. During the last years of the nineteenth century and the first three decades of the twentieth century, Freud singlehandedly developed what some labeled and some continue to describe as the pseudoscience of psychoanalysis, with its emphasis on psychological rather than physical causes of mental illness.

Today, Freud and Alzheimer represent the opposite sides of a battle that never needed to be fought. We now know that psychological factors affect brain function, while, in turn, brain dysfunction influences such expressions of mind as perception, judgment, thought, and mood. Not surprisingly these diametrically opposed interpretations of mental illness created tensions over the years that on occasion broke out into outright acrimony.

Up until the 1970s American psychiatry was dominated by psychoanalytic or other psychologically based theories of mental illness. On the other side of the aisle, aspiring neurologists evidenced little interest in social, cultural, or psychological components of the brain diseases that they studied. From the 1980s onward, this split between brain-based and psychological culture-based theories of disordered thinking continued. During this time, several facts were either ignored or actively suppressed.

Emil Kraepelin, who identified schizophrenia, believed *Dementia praecox* resulted from dysfunctions of the frontal and temporal areas of the brain. Since there was no treatment for these dysfunctions, it was expected that there would be no improvement. But when some of the patients with *Dementia praecox* improved, doubt was cast on schizophrenia as a form of dementia. Investigators failed to notice or neglected to mention that many cases of schizophrenia—as with *Dementia praecox*—failed to recover.

Not until 2022 was the implication of this observation provided by a brilliant neuroscientist Nikolaos Koutsouleris from the Max Plank Institute in Munich, Germany:

Our studies revealed that young patients with poor functional outcomes [from schizophrenia] overexpress brain patterns over time in line with Kraepelin's concept of dementia praecox as a progressive frontotemporal disorder.

Whatever Alzheimer may have been thinking while on the night train to Breslau, his thoughts were interrupted by feelings of physical discomfort. Within hours of arrival, he had spiked a temperature and was admitted to a hospital. The illness, believed now to be streptococcal infection, led to rheumatic fever and kidney failure. He died several days later in Breslau at age 51.

WHO IS CRAZY NOW?

Consider these two people:

A 25-year-old man voices delusions, experiences hallucinations, and cuts off all communications with relatives and friends. His diagnosis is schizophrenia.

An older person, age 65, also displays delusions and agitation for the first time in his life; he progresses to specific symptoms known to be associated with a particular degenerative brain disease, such as the loss of memory in Alzheimer's disease.

Throughout most of the twentieth century, these two profiles were considered wholly distinct from each other and even cared for by different specialists: neurologists for dementia and psychiatrists for schizophrenia and other *functional disorders* (illnesses thought to be unrelated to the brain). Yet, one has only to revisit the writings of the early researchers who first recognized these disorders to appreciate that they considered them closely related. Even today, it can sometimes be difficult to distinguish them, especially if the illness occurs in early middle age or younger. Thus, a slowly developing case of Alzheimer's or other dementia may initially be cared for by a psychiatrist. At the other extreme, someone with a profound depression will perform poorly on neuropsychological testing as a result of the general slowing caused by that depression. As a result he may be diagnosed with dementia and come under the care of a neurologist. Not until an astute physician of either specialty recognizes and treats the depression will the patient return to health via a nearly coincidental improvement in both depression and neuropsychological test performance.

But one does not have to go very far back in time to encounter tensions regarding psychiatric and neurologic diagnoses. During my years in medical school, diseases of the mind were rather arbitrarily distinguished from diseases of the brain. Later as a psychiatric resident, I frequently encountered incredulity and resistance from many of my psychiatric teachers whenever I suggested a particular behavior disorder might be secondary to a brain dysfunction. One of them went so far as to say to me, "The problem is that instead of caring for your patients by talking with them about their unconscious drives and their interpersonal conflicts with others, you go right along with their belief that their problems stem from something affecting their brain. If you are so convinced of this belief, maybe you might be happier as a neurologist."

Unfortunately after I switched to neurology, I encountered a similarly frustrating but opposite situation. Neurologists at that time had little curiosity or interest in any behavioral issues in their patients. These were best left to psychiatrists, "the shrinks," as they were fond of referring to them. Neurologists were much more comfortable dealing with movements and paralysis, and perception and its absence (loss of appreciation of sensation secondary to a stroke, for example). This dichotomous relation between the "shrinks" and the "reflex hammerers" (a down-putting reference to neurologists, who use hammer-like, rubber-capped instruments to test reflexes by tapping muscles) became unsustainable in the last two decades of the twentieth century and the first two decades of the twenty-first. Fortunately, the numbers of these mistaken diagnoses are decreasing with the burgeoning of the hybrid specialty of neuropsychiatry.

"I AM JESUS CHRIST"

What caused this revolutionary reappraisal whereby psychiatric illnesses (also referred to as *functional illnesses*) were recognized as resulting from brain disease? Briefly, patients were discovered who showed signs of schizophrenia caused by epilepsy. As an example, for over twenty years, I've treated a man, who was originally diagnosed with schizophrenia, with an anticonvulsant. He first came to my attention when he was brought to the emergency room after making a disturbing claim. While stalled in a traffic jam an hour earlier, he suddenly opened his car door and stepped out onto the highway. To each person who rolled down their car window, he repeated the claim, "I am Jesus Christ." Not surprisingly, it didn't take long before police were on the scene, and he was escorted to the local emergency room.

He was in his mid-twenties, living with his parents, jobless—he seemed like a textbook case of schizophrenia. But when his sister told me of occasional staring and attentional lapses that she had observed in her brother, I ordered an electroencephalogram, the so-called brain-wave test, which turned up positive evidence of epilepsy. Since the highway incident wasn't the first example of a public claim of messianic identity, previous psychiatrists had ordered antipsychotic medication, to no avail. He either didn't take them or, if he took them, they didn't seem to be of any help. Based on the sister's observation and the results of the electroencephalogram, I replaced his antipsychotic with an anticonvulsant. The result was dramatic. Within days his delusion disappeared, and he regained insight into the delusional nature of his previous claims. Over the next twenty years under my care, he has remained free of delusions and works regularly as a self-employed handyman.

As another example, neurosurgeons and neurologists also discovered that electrical stimulation of the brain, especially of the temporal lobes, elicited what one neurosurgeon referred to as an "experiential illusion": the patient, without losing his or her sense of being on the operating table while undergoing such treatment reported the sensation of traveling back many years to recall, reexperience, and describe events from childhood. This wasn't a delusion or a sign of mental instability, but rather an actual physical effect of the stimulation of a specific area of the brain.

But the most striking alteration in our understanding of a specific disease occurred with Parkinson's disease. The first description of the shaking palsy dates to 1877 and physician James Parkinson. Parkinson and subsequent neurologists described tremor at rest in the hand (or hands), stiffness in the arms and legs and trunk, slowing of movement, and poor balance. Both the senses and the intellect were described as "unaffected," as Parkinson wrote. This remained the accepted view well into the late twentieth century.

"Intellectual capacity may continue unimpaired" wrote a famed and knighted British neurologist with the quintessentially eponymous name for a neurologist—Lord Brain. Notice that all the symptoms mentioned so far involve movement (the motor system) and hence land in the bailiwick of neurologists. During my psychiatric residency, I don't remember seeing even a single example of Parkinson's disease.

Further observation over many years of people afflicted with Parkinson's disease showed that if observed over a sufficient time frame, the course of the disease often changes dramatically. About two years after diagnosis, many of the patients began showing problems with attention, memory, language, and organization. Accompanying these impairments came more psychiatric-appearing symptoms, such as apathy (the patient doesn't seem to care that he or she is getting sicker), depression, delusion, and sleep difficulties, including daytime sleepiness. In fact, neuropsychiatric symptoms typically precede the onset of motor symptoms; cognitive changes can also be demonstrated even in newly diagnosed individuals with Parkinson's disease.

Currently, Parkinson's dementia and Lewy body dementia are common causes of dementia after Alzheimer's disease. Unfortunately, three-quarters of people with Parkinson's disease will develop dementia within ten years after receiving their diagnosis. In short, Parkinson's disease is now recognized as showing behavioral disturbances that require the same medications used for diseases considered primarily psychiatric. We encounter here once again the logical peril of assuming that illnesses of the mind can be parceled into narrow categories of strictly brain disease to be treated by neurologists or functional disorders to be placed under the purview of psychiatrists.

A QUESTION TO PONDER

Would you prefer remembering a lot of things that never really happened or not remembering a lot of things that did?

Obviously, the correct answer to this somewhat enigmatic riddle is neither. But read the alternatives again. The first implies a serious disturbance in thinking, perhaps a psychosis or at least some form of mental imbalance. The second alternative refers to memory alone and is compatible with the early stages of Alzheimer's and other dementias. But actually, either of these unhappy alternatives could define an illness fitting either a psychiatric or neurologic disease. It would all depend on when the patient was seen, along with the trajectory of the illness and the practitioner.

When I primarily practiced psychiatry, I encountered patients remembering things that hadn't happened. Rather than suffering from illnesses of psychiatric origin, they represented the early stages of Alzheimer's, with the affected person making his or her best attempt to make up for lacunae in their memories by inventing imaginative scenarios. Thus they represented a brain disease and not a primarily psychiatric illness at all.

Now when I'm primarily engaged with neurologic patients, it is not at all unusual to find someone with a faulty memory but who is actually suffering from depression; once the depression is successfully treated, the patient will no longer display memory disturbances.

Distinctions between brain-related illnesses and others inaccurately thought to somehow be free from any dependence on the brain are wrong. Most of them exist along a continuum from mild to the need for professional help (often medications), or in the worst instances, hospitalization. Psychiatrists and neurologists are taking on a totally new identity in the twenty-first century, and the two fields appear to be merging into the hybrid identity of a neuropsychiatrist—a doctor with full training in one field (neurology or psychiatry) and extensive training in the other. From this vantage point, it does not seem as strange as it did in the late twentieth century to think of Alois Alzheimer as a psychiatrist or of Sigmund Freud as a neurologist who in his midcareer turned his attention to matters more typical of a psychiatrist.

Here is a trick question given to neurologists and psychiatrists when qualifying for their board examinations: "Who was the nineteenth-century neurologist who early in his career wrote authoritative textbooks on cerebral palsy and aphasia?" In most cases, the aspiring candidates for board certification in either neurology or psychiatry miss this question unless they are aware that Sigmund Freud began his career as a neurologist.

CHAPTER III

DEMENTIA: HOW IS IT CONFIRMED?

EVERYTHING YOU NEED TO KNOW ABOUT THE BRAIN TO UNDERSTAND DEMENTIA

Earlier I suggested four questions to screen for the likelihood of Alzheimer's or other dementias. What comes next?

Whenever Alzheimer's or other dementias are suspected, the investigation follows the following sequence:

First a **neurological examination** is carried out, looking for signs of brain damage. These can include localized weakness, clumsiness, and a vast array of indicators of brain dysfunction. Particularly important is the presence of so-called release signs. As the brain matures, certain signs observable in infancy disappear, only to be replaced by more sophisticated responses. For example, infants cannot control their individual fingers until certain nerve tracts mature. At maturation, the so-called primitive reflexes disappear. Later in life, in response to brain damage, these primitive reflexes reappear. One test for this is performed by rubbing a smooth, nonpainful instrument like a pen cap across one of the palms. In the case of brain damage to the frontal lobes, this stimulus brings about a brief twitch in the muscles of the chin. Nobody is entirely certain about the basis for this *palmomental reflex*. But it is clearly abnormal and a reliable sign of frontal brain damage.

Equally important as the neurological exam are neuropsycho-logical tests that measure the brain's cognitive functions. Although most cognitive functions interrelate with each other, certain ones serve as the entry point for evaluating the others. Put somewhat differently, if these functions are not evaluated first, nothing very meaningful can be said about the others. Absent alertness and attention, one cannot evaluate memory, for example.

Here is a diagram of the different cognitive functions. Not all of them need to be tested every time, depending on the problem being evaluated. Let's take language as an example. If language is the problem (difficulty finding words, for instance), then the affected person is asked, "What do you call something that uses ink and you write with?" (*naming to definition*). Next he is shown a pen and asked to name it (*confrontation naming*). Next, he may be instructed, "If today is Tuesday, please close your eyes for a moment" (*comprehension*); "How do you spell the word automa-tion?" (*spelling*); "Please read these first three sentences from this magazine article" (*reading*); "Repeat after me 'no ifs, ands, or buts'" (*repetition*).

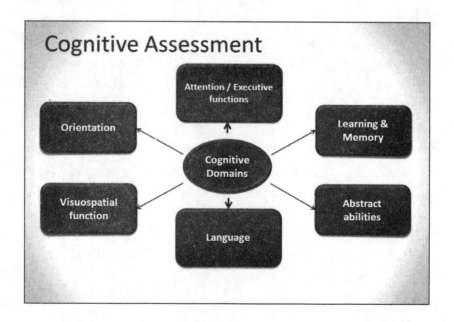

Here is a description of the tests used to assess the different cognitive functions:

Orientation: Orientation is the first function to be evaluated. Person, place, time—these three pieces of information provide a broad assessment of a person's functioning. Does the person know the year, month, date, season, day of the week, as well as the approximate time without looking at a watch or clock?

Attention: Attention is always dependent on arousal (alertness) and motivation (willingness to cooperate). If asked to define attention, I'd favor "the ability to focus on one thing to the exclusion of others." It is tested by asking the person to repeat back a series of numbers in both the forward and backward direction. Next, subtract 7 from 100 and then 7 from that number, and then 7 from the next number. Ask the person to continue to 65 (100, 93, 86, 79, 72, 65). Attention and executive function are frequently investigated together because they serve as probes for the integrity of the frontal lobes (to be discussed in detail in Chapter VI). One of the frontal-lobe functions is inhibition. To test that, ask the patient the following: "I'm going to recite a series of letters. Whenever I say the letter A, put up your right hand; for all the other letters, raise your left hand." Such a test requires the active inhibition of mistakenly raising the left hand when hearing the letter A.

Divided Attention: Such a test measures ability to switch attention from one task to another. "Name the last five presidents." Upon completion, say, "Now recite those presidents in alphabetical order" (also an excellent test of working memory—moving information around without writing anything down).

Memory: Request that the person repeat back a list of five words after hearing them spoken.

Delayed Memory: Five minutes later, ask him or her to recall the list of words once again.

Naming: Ask him or her to identify the names of animals from looking at sketches of the animals.

Language: Administer tests gauging the ability to understand a series of sentences and repeat them. "Mary told John that he was the person Sean named as the leader of the club. Who was chosen as a leader?" Next, the person is asked to "Name as many animals as you can in one minute." As an alternative, the list can involve words instead of animals. "Name as many words as you can beginning with the letter *F.*" Most people can name at least eleven words or animals in one minute.

Abstraction: Ask the question "What do an apple and a cherry have in common? Answer: They are both fruits." Although the answer is frequently given, "They are both round," that is not the best answer. Fruit represent a higher level of abstraction.

Visuospatial: Request that the person draw a clock face with the hands at ten minutes past eleven. Also have the person draw a three-dimensional cube.

Based on the neurological and the screening neuropsychological exam, all responses and findings suggestive of dementia will be selected for a more thorough investigation. Why? People with normal mental functioning and those with very mild cognitive impairment (MCI) may escape detection at this screening level of testing.

Next, **brain imaging** will be employed. One or all of the tests taken together can detect strokes and tumors and other structural causes of dementia. Included among the brain-imaging studies are *Computed Tomography* (CT), which uses X-rays aided by computers to produce brain images, and *Magnetic Resonance Imaging* (MRI), which uses magnetic fields and radio waves to construct more highly detailed brain images.

Genetic Tests: Some cases of Alzheimer's (primarily early onset) and the other dementias are inherited. A series of genetic tests can determine if a person is at genetic risk for dementia. These tests are not by themselves sufficient to tell someone who tests positive that he or she has the disease, or even that the person will eventually come down with it. This important point will be elaborated upon at several places in this book.

Finally, blood tests are now available (and in active use) to measure the levels of beta amyloid and tau protein in people suspected of Alzheimer's.

LOCALIZATION VERSUS HOLISM: THE SEARCH FOR A COMPREHENSIVE UNDERSTANDING OF BRAIN FUNCTION

Alzheimer's and the other dementias are the behavioral and cognitive expressions of brain disease. But where in the brain?

Over the past half century our understanding of the brain has undergone a profound revolution. For the longest time (well into the mid-twentieth century) the emphasis was on *localization*: the belief that human functions along with behaviors and propensities were the products of specific brain areas. Taken to the extreme, localization theory led to the pseudoscience of phrenology.

For instance, Franz Josef Gall, the most famous proponent of phrenology, believed that the brain was composed of many "faculties" located at various points that could be identified by touch on the surface of the brain. These brain areas influenced the formation of the skull. Gall believed those areas that were the biggest tended to form prominences on the external surface of the skull. By palpating the skull, the phrenologists claimed to be able to gauge a person's capacity for love, mirth, order, sublimity, and so on. Today neuroscientists recognize all of this as nonsense.

While it's true that simpler actions, say, the movement of an arm, may be fairly well localized to a specific motor area on the opposite side of the brain, no such localization is possible for higher-order faculties (love, courage, religion, conviction, etc.). Indeed, the very word *faculty* is rarely used now. *Power* is the usual substitute: power of speech, power of reasoning, etc.

The more contemporary view of brain function is based on *holism*: the belief that the brain functions as an interconnected whole. According to this interpretation, every neuron has the potential to influence the actions of every other neuron. Here is a diagram of a simple circuit.

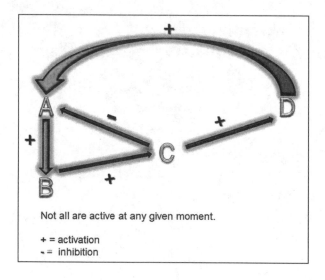

Not all are active at any given moment.

+ = activation
- = inhibition

A positively influences B which causes B to fire. B then stimulates C, which then influences both A (inhibiting its ability to fire) and D, which then completes the circuit by activating A and overcoming the inhibition exerted by C.

Now take a simple circuit like that and extend it to the 86 billion neurons in the human brain. The results are sufficiently complicated that understanding the brain's functioning requires the help of powerful computers buttressed by AI programs.

Fortunately, an easier approach is available that will serve our purposes quite nicely. Dementias can be understood by concentrating on the functions of the frontal and temporal lobes along with their immediate connections. If you remember that information, you are well on the way to a slightly simplified but adequate understanding of the dementias.

First, I'll cover the **temporal lobe**, located on the side of the brain and corresponding roughly to the area behind the ears when seen from the side. The temporal lobe connections include the seahorse-shaped hippocampus, which initially registers our memories.

After formation, a memory is distributed from the hippocampus to various points of the cerebral cortex. But the process also works backward—the cerebral cortex messaging the hippocampus—when we retrieve that memory. It is also important to mention that each retrieval of a memory causes subtle differences when the memory is restored. The chemical structure of the recalled memory often differs slightly from the chemical structure of the original memory. Thus, memories are not like recordings or DVDs but have a dynamism about them that can lead to later faulty recall of an event. The temporal lobe is also important in self-recognition and identity. If the temporal lobe is electrically stimulated at surgery, it's possible to retrieve highly complex memories of past experiences, so called *experiential illusions*.

Several of the dementias I will describe involve visual hallucinations, often of tiny people, friendly animals, and familiar scenes. Think of these as originating in the temporal lobe and its connections. Loss of memory—the hallmark of Alzheimer's—is linked to the damage or destruction of the hippocampus (principally the dentate gyrus, a part of the hippocampus). In fact, mild cases of Alzheimer's may involve memory difficulties only. But if asked about additional symptoms by the neuropsychiatrist, the Alzheimer's person may describe a loss of sense of self. As you recall, Alzheimer's original patient described just such an experience "I've lost myself, so to say," was Auguste Deter's frequent response to Alzheimer's questions.

In a PET scan image, the temporal lobe almost always shows profusion (flow of blood or glucose) abnormalities in Alzheimer's disease. Before we move on to the frontal lobe, the second most important brain area involved in dementia, let's describe memory in more detail.

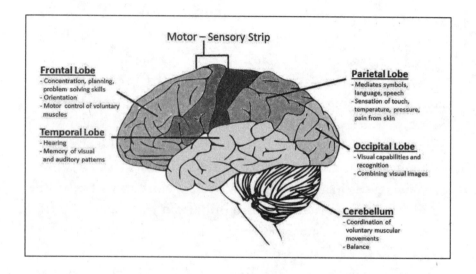

CHAPTER IV

THE TEMPORAL LOBES AND MEMORY

A TRIP DOWN MEMORY LANE

Without our memory, who are we, really? If we are unable to recall our life's events, the people we have known, our thoughts, feelings, and behaviors from the past, we are lost. Memory forms the basis for our identity. The seventeenth-century philosopher John Locke even defined identity in terms of memory. But when we think about memory, we must specify what type of memory we are speaking about. Memories—like dogs, cars, and plants—come in several varieties.

- **Episodic memory.** As the word implies, episodic memory refers to specific events. I'm writing this sentence *now*. I'm experiencing this as a discreet episode. Later I may not recall the exact time or circumstances of the sentence's composition. At that time, memory enters the realm of semantic memory, shorn of all the details regarding time, place, etc.
- **Semantic memory.** Most memory is stored in semantic memory. We may know that Charles Dickens is the author of *A Tale of Two Cities*, but it is unlikely we can remember the exact moment we learned this fact. If we can remember the occasion, the circumstances most likely involved something emotionally arousing. Perhaps we were asked

to read a passage from the novel before our whole seventh grade class. At that moment, the brain's emotional center, the amygdala, became activated and placed a kind of emotional valance (of anxiety) on the experience. As a result, we remember with great clarity this reading-to-the-class episode. While episodic memory always requires consciousness, our huge repertoire of semantic memory only enters our consciousness when we are prompted, usually by a request. (Who wrote *A Tale of Two Cities*?)

- **Working memory.** This third variant of memory is the most interesting and the most challenging. Your facility with it is probably the single greatest indicator of your overall intelligence. Further, working memory is the most important variant of memory to augment by practice. This involves taking several items and manipulating them in your mind. Tell me the players on your favorite baseball, football, or soccer team. Now without writing anything down, list the names in alphabetical order. Next—and hardest—list the names in order from highest to lowest according to the number of letters in the name. To accomplish any of these three requests, you must envision the players and the players' names and then shift them around in your mind according to the instruction. Try it. Not so easy, is it? In fact, don't be discouraged if you can't comply with that last request. You really must possess a superpower, highly honed working memory to carry out such a highly demanding and complex task.

- **Procedural memory.** This is a memory that cannot be put into words. As I write this sentence, I control my pen (I better, if I expect to read any of this later), but I can't explain or inwardly experience *how* I do it. The muscles in my hands and forearm automatically come into synchrony, and the writing process proceeds smoothly. Riding a bicycle, driving a car, and responding in racquetball—each performance is initially learned by practice, which requires conscious awareness. Later, the performance becomes automated—best not to think too consciously about it lest we interfere with a smoothly automated performance.

Notice that two of the forms of memory I have discussed so far, episodic and working, require conscious awareness. Semantic memory and procedural memory, in contrast, reside outside of conscious awareness in our brain and consist of all the knowledge that we have ever accumulated. Only by converting semantic memory into conscious awareness of episodes (episodic memory) are we aware of its contents.

Here is a working diagram of all of this. Let's peruse it starting at the upper left-hand corner.

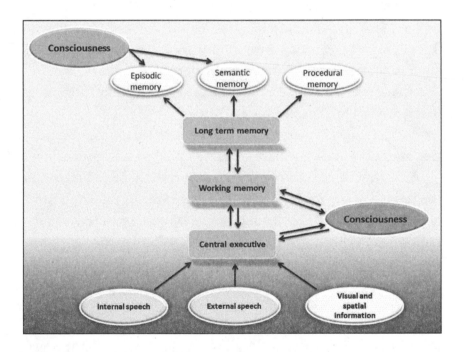

Consciousness accompanies **episodic memory** (always) and **semantic memory** (only at the moment of bringing information into conscious awareness, which is often in response to a request).

Procedural memory does not involve consciousness, which only interferes with its smooth operation. Athletes speak of "choking" or "clutching" when, under pressure, they attempt to alter deeply ingrained memory responses.

All three forms of memory (episodic, semantic, procedural) converge into long-term memory.

Now look at the bottom of the diagram:

- **Internal speech:** Silent internally rehearsed self-talk, as with a chess player pondering his response to an opponent: "If he moves that bishop, I'll get him!"
- **External speech:** Talking and responding to other people talking.
- **Visual and spatial information:** Using information gathered by the senses.

These three sources of memory converge on the central executive, another term for the self-experiencer of consciousness.

Finally, all of the above sources have a direct entry to working memory and consciousness.

As a means of breathing life into these terms, consider this example:

As I'm writing, I can hear in the periphery of my attention the sounds of my wife one floor below me. She is on Zoom with her French teacher. Her learning of vocabulary, syntax, and pronunciation will call upon *episodic memory* (this particular lesson is taking place now), *internal speech* (reviewing the words), and *external speech* (pronouncing the words). Added to this is visual information (the book Carolyn and the teacher are reading, *The Little Prince*). As each word or phrase is learned, it is stored in both episodic (initially) and later semantic memory. When she is asked to summarize what she has read, each muscle of articulation along with the muscles of the diaphragm and chest mold themselves into specific confirmations corresponding to the sounds of the words in that language. If she had learned French earlier in life, the muscles would now be functioning flawlessly—the throat muscles perfectly coordinated to pronounce French words without any suggestion of an accent. Such coordination isn't done consciously but via procedural memory, sometimes mistakenly dubbed "muscle memory." Muscles do not possess memories. Rather, they are brought into synchrony by the nerves, which tune them like a musical conductor, coordinating the different instruments. Gesture, loudness, and flexion will gradually become part of Carolyn's procedural memory.

If you've ever studied a foreign language, think back to the earliest lessons. If your teacher was enthusiastic, you can probably remember those earliest lessons (episodic memory). Gradually those lessons shifted into semantic and procedural memory. The goal was an instantaneous familiarity with a new language via combination of internal speech and external speech. Once you achieved a certain fluency, you no longer had to think about or monitor your speech. The whole process occurred as part of procedural memory. Only when you encountered an unknown word or idiom did you have to pause and consciously search for the word's meaning (internal speech and semantic memory).

Where does working memory fit into all of this? It is also keeping several items simultaneously in mind and shifting them around according to request or circumstances. The only difference is that in this case—Carolyn's French lesson —the items are in a language other than English.

Knowing these different varieties of memory sometimes enables us to come up with explanations for the seemingly inexplicable.

DON'T CLOSE THE GARAGE DOOR

This morning I walked our dog, Leah, a bit earlier than usual. As I was leaving the house, I said to my wife, "Don't close the garage door when you go to meet with your friends." Forty-five minutes later, Leah and I returned from our walk only to encounter—you guessed it!—a closed garage door.

While a therapist would probably ask, "Was your wife annoyed at you for some reason?" or a neurologist might ask, "Could this be an early sign of dementia?" I don't think either of these questions get to the heart of the matter or provide any explanation why I found myself standing in the early morning January chill wondering what to do next. (I don't ordinarily carry my cell phone on these walks.) The most likely culprit here, I believe, is the overpowering influence of procedural memory—the type of long-term memory that involves carrying out different skills and actions without processing them through conscious awareness.

After several years of driving, most people no longer need to think about the mechanical aspects of driving. The driver doesn't have to look down at the gear indicator to know whether the car is in neutral, drive, or reverse. The driver can "feel it." Thanks to long-established habit patterns, driving can be carried out with a minimum of conscious effort via procedural memory.

When my wife got into the car, her procedural memory clicked into place and overrode her episodic memory established fifteen or twenty minutes earlier when I had asked her specifically not to close the garage door. Sometimes the consequences of a mix-up involving different types of memory can be more than just annoying; they can be tragic, such as when a parent driving his or her infant in the backseat of the car forgets about the tiny passenger upon arrival at the destination. The parent returns to the car hours later to discover the child dead in the car seat. This horrific experience has a name: forgotten baby syndrome (FBS). We all have to be on guard to prevent deeply rooted procedural memories from overriding our episodic and working memory.

When I asked my wife what she had been thinking about when she was backing out of the garage, she replied, "You always take the dog out after I leave. So, I assumed you were still in the house"—an interesting and, I don't doubt, truthful account. Her experience-based assumption that I was home, plus the activation of procedural memory for driving, exerted a greater influence on her behavior than my clearly articulated request earlier not to close the garage door.

It is useful to remember that "I forgot" sometimes involves, as in this example, a shifting of memory processing, rather than an actual failure of memory powers. That distinction is important, because when you say, "I forgot," you risk being on the receiving end of other people's fury. These two words can get you fired, deliver a setback to your marriage, or ruin a friendship. But "I forgot" can result from a panoply of causes that often have nothing to do with a faulty memory or incipient dementia. They can be based on using the wrong type of memory, rather than an actual loss of memory.

From a practical point of view, therefore, we all have to be on alert whenever we are carving out an exception to our daily routine (episodic memory), lest the well-established procedural memory automatically take over, resulting in inconvenience or even disaster.

AFTERNOON AT THE FAIR

You are vacationing on Martha's Vineyard and decide to attend the Agricultural Society Livestock Show and Fair. Since you are from a city, you have little experience with farm animals and feel tense when you're around them. At the livestock section of the fair, you encounter a tamed goat and are thrilled to watch the goat eating grass and hay from your hand. As a result, whenever you smell grass in the future, you may experience the same comfortable laid-back feeling that you felt that afternoon at the fair.

Now let's consider the same circumstances, with you participating in a different scenario. In this instance, the goat is unfriendly and tries to nip your hand when you attempt to feed it.

If you were almost bitten, the smell of fresh grass or even a picture or video of a goat may elicit tension.

Ever afterward, these two experiences will be remembered differently. Depending on your experiences with the goat, your brain will either remember the interaction as fun (a positive valance based on the memory of the gentle goat) or as a close call that conceivably could have resulted in the savaging of several of your fingers (a negative valance to the memory). All of this makes perfect sense—good or bad experiences will inevitably lead to memories with a corresponding valance. If any experience is hurtful or upsetting in any way, your memory for it is likely to be disturbing as well. Surprisingly, such a vast difference in memory for different experiences depends on only a single tiny molecule.

Neurotensin is one of more than twenty-five biologically active peptides (short strings of amino acids that form the building blocks of proteins) found throughout the brain.

Here is what happened during the two hypothetical visits at the Agricultural Society Fair. The brain responded to the two experiences at the moment, and neurons released neurotensin so as to shift the experiences (the docile goat versus the hostile one) along different neuronal pathways where they are encoded in the form of positive or negative memories.

Which of the two emotional valances do you think has the more commanding effect on behavior?

If your world view tends towards pessimism, you will not be surprised (or disappointed) to learn that our brains are hardwired to preferentially respond to negative experiences. Your financial advisor may have that fact in mind when advising against checking your portfolio too frequently. According to the principal of *myopic loss aversion*, the more often you check how your financial account is doing, the more likely you are to spot a loss, and since losses affect the brain more than gains, you are more likely to buy or sell, churning turnover. This is just the opposite of how you should be responding according to a classic study "Trading

Is Hazardous to Your Wealth," by professors Brad Barber and Terrance Odean.

But back to the fair. If that goat had actually bitten you, the odds are you would for good reason remember that experience with a permanent negative valance. In the positive scenario where the experience went well (no biting) there was no particular reason for you to experience future anxiety. As a result, the experience will be less clearly etched in your brain.

Memories that link unrelated ideas—like "the fair," "goat," "grass feeding"—are referred to as *associative memories,* and they are frequently emotionally charged. These emotions are formulated within the tiny almond-shaped regions on each side of the brain referred to as the amygdalae (amygdala in the singular form). Although the amygdala is rightly known as the brain's "fear center," the structure also responds to pleasure and other positive emotions.

Memory ability is perhaps the most important contributor circumventing or preventing dementia. The easiest way to enhance memory is to exercise it. In Chapter X we'll discuss a program for that purpose that can and should be used daily. But first let's take up the second of the two brain areas most important in understanding dementia.

CHAPTER V

THE FRONTAL LOBES AND THINKING

"MY WIFE IS A DIFFERENT PERSON"

The frontal lobes are located just behind the forehead. If you touch your forehead with the tips of your second and third fingers and then slide your fingers down to the bridge of your nose you've outlined the anterior and inferior dimensions of the frontal lobes, one on each side of the brain. The best way of understanding what the frontal lobes do is to observe the consequences of frontal-lobe injury.

Let me introduce you to M.S., a real patient of a colleague. By 45 years of age, she had established herself as a successful businesswoman in a large company. One day at work, she experienced an epileptic seizure—the first in her life. A CAT scan revealed a tumor in the right frontal lobe. The tumor required surgery, and the operation went smoothly. The seizures ceased, and she recuperated at home.

At a check-up several months after the surgery, she reported no problems. Her husband, however, reported several problems. Formerly a lifetime early riser, his wife slept late and once out of bed took an inordinate amount of time showering, dressing, and getting her things together. On a typical morning her preparations delayed her to the point that she rarely made it to the office before 11:00 a.m.

While at her desk at work, she was easily distracted; she tended to shift from one project to another without completing any of them. Her work and home behavior decompensated to the extent that both her boss and her husband described her as "a different person."

Reluctantly, she agreed to psychological testing. Her IQ remained above average. Her greatest difficulty involved organizing, scheduling, and switching tasks. She knew what had to be done, expressed a willingness to do it, yet could not coordinate the various components of a task to achieve what she wanted and what others expected of her.

Throughout all this she knew, thanks to her neurologist, what her problem was: impairment of the frontal lobe secondary to the tumor and the consequent surgery. But she still insisted that she could perform as well as previously. It was as if at one level she recognized her difficulties—even occasionally commenting that she might not recover sufficiently to do her work—while at another level she denied the problems altogether. This was especially puzzling since she achieved a normal-to-superior performance on neuropsychological tests, including memory.

Another peculiar aspect of her behavior: She had few problems with routine tasks but failed miserably when she had to come up with a novel response. She was at her worst when she had to focus attention, abstract necessary information from background material, produce plans of action, respond with flexibility to changing circumstances, or evaluate the outcome of her actions. All of these were the expected functions of a high-level executive, but she was unable to do them or demonstrated impairment in performing them. The reason? These operations are carried out by the frontal lobes, the brain area destroyed by her tumor.

As a result of her tumor, M.S. was impaired in all five of the major frontal lobe functions:

- **Drive/motivation.** Ambition and self-motivated behavior are usually the first casualties from frontal-lobe damage. M.S., formerly an ambitious and motivated worker, started sleeping late after her injury. She required frequent reminding and prodding by her husband to keep her on schedule. At work she needed monitoring and supervising to keep to her work organized.

- **Sequencing.** M.S. experienced difficulties keeping information in proper sequence and within context. Unable to monitor her own performance, she frequently shifted from one project to another—hardly starting one before turning her scattered attention to another. Unable to integrate several sources of information (multitasking), she returned again and again to previously reviewed material in a failed attempt to integrate it.

- **Executive control.** This involves keeping the big picture in mind. Think of an executive of a moderately sized corporation. He or she must balance many competing interests to make the corporation perform at maximal efficiency. Executive control applies on an individual level as well. "What should I be doing now that will enable me to work most efficiently?" M.S was unable to answer this, since she had lost this frontal-lobe function.

- **Advanced planning.** In order to plan ahead, it is necessary to be able to compare how things are with how one wishes them to be. The resulting imaginative model of the future serves as a guide for alternating and updating one's behavior in the service of reaching that goal. With frontal-lobe damage, a stable internal model of the future cannot be maintained over time.

- **Self-analysis.** "Self-awareness, consciousness, or self-reflectiveness is the highest psychological attribute of the frontal lobes," claimed Donald T. Stuss, one of the world's most lauded experts on the frontal lobes. M.S. totally lacked analytical powers, shifting her mind from one time to the next on the subject of whether she had recovered enough to continue her work. When this question was presented to her in an objective problem-solving perspective, she admitted she couldn't do the work—and then, sometimes within minutes, she reversed herself. She lacked *insight* into her impairment and its effects.

If we wish to summarize frontal-lobe function in one word, that would be *thinking*—clear, cogent, logical thinking. All of the dementias are characterized by thinking disorders in one or more of the five frontal-lobe functions mentioned. Moreover, just as we can strengthen our memories by memory exercises, we can strengthen our thinking powers by exercises in logic, reasoning, and abstraction. When we regularly carry out these exercises, we strengthen our thinking powers and help avoid dementia.

BLACK SWANS AND SAMPLING ERRORS

Thinking cannot occur without a topic—our thinking is always about something. We choose (or other people or situations force us to choose) the things we think about. Focused attention on a subject distinguishes thinking from daydreaming.

While many different criteria are essential for effective thinking, several are especially important: logic, precision, significance, scope, and—most importantly—the ability and willingness to evaluate the qualities of one's own thinking.

Traditionally, thinking is divided into *deduction* and *induction*. Deduction begins with a general rule and proceeds to specific examples: "All dogs are animals. Leah is a dog. Therefore, Leah is an animal." Induction involves generalizing from several examples to a general rule. But we must remember that generalizations resulting from my induction may be based on insufficient sampling. No matter how many white swans were observed, it could not be concluded that black swans didn't exist. Only when this sampling error was overcome by more observations of swans was the existence of a black swan verified.

A FISH CANNOT EAT A CROW

In the early part of the twentieth century, the Swiss psychologist Jean Piaget first postulated that thinking starts in young childhood and progresses until about age twelve, when the child enters what he termed the formal operational stage. At this stage of development, the child can think in abstract terms, follow logical prepositions, and reason via hypothesis. Piaget's emphasis was on the importance of genetics and maturation. To summarize his work: as we grow older, our power of abstraction increases.

A contemporary of Piaget, the Soviet psychologist Alexander Luria, investigated the importance of cultural rather than strictly genetic or maturational factors in thinking. His research was carried out in 1931 among the inhabitants of a remote Russian village, prior to the introduction of widespread comparative modernization in Russia. Luria's subjects, who could neither read nor write, spent their lives working on cotton farms and based their thinking on personal experiences rather than what we would call logical thinking based on abstractions.

Luria described the thinking processes: "In this mode of thought the primary function of language is not to formulate abstractions or generalizations, but to revive practical situations."

Luria provided an example of one subject's response to the question "What do a fish and a crow have in common?" Instead of responding, "They are both animals," as most people would today, he emphasized the differences: "A fish lives in the water. A crow flies. If the fish just lies on top of the water, the crow could peck at it. A crow can eat a fish, but a fish cannot eat a crow."

For Luria's subjects, abstraction and generalization were difficult to appreciate or express. But with greater education, the shift from what Luria referred to as primitive thinking to abstract thinking increased. In the late twentieth century, the psychologist James R. Flynn discovered that the ability to think abstractly is associated with an increase in generalized IQ.

But failures in abstract thinking are not a thing of the past. On the Montreal Cognitive Assessment (MOCA), the gold standard in assessing early dementia, there is a test for abstraction. Examples of similarities are provided such as a banana and an orange, and the subject is asked to come up with the similarity between them (in this case, they are both fruits). The test-taker is then asked the similarities between a bicycle and a train, and a watch and a ruler. The train and bicycle similarity, of course, is that both are means of transportation, and the watch-ruler similarity involves measurement.

Not everyone within the general population is capable of coming up with a correct response to the bicycle-train and watch-ruler similarities. A certain percentage of people without neurologic or psychiatric illness will come up with something like this: "A train and a bicycle both have wheels"; "A watch is worn on the wrist, or a ruler could be used in geometry class to draw angles." Both answers display little evidence of abstraction. More advanced abstraction impairment leads to concrete responses where any functional similarities disappear altogether: "The watch and the ruler can both be made of steel"; "The train can run over a bicycle on the tracks and the bicycle can run into the train at the crossing." Concrete thinking is characteristic of schizophrenia and many of the degenerative brain diseases such as Alzheimer's. But it also occurs in normal people with reduced intelligence and education.

Concrete rather than abstract thinking is only one example confirming that a thinking disorder does not necessarily imply mental abnormality or deterioration but exists on a continuum, with the final result depending on one's intelligence and education.

Other thinking oddities take the form of thoughts that sound illogical and even bizarre on first hearing but make sense when put into a special context. "Should I take the bus today or pack my lunch?" a husband asks his wife before heading off to work. Such a question would be frustratingly opaque, unless his wife remembers that on rainy days her husband drives his car to work and eats a packed lunch in the office, but on clear days he takes the bus and walks to lunch at a nearby restaurant. In this example, anyone unfamiliar with the situation cannot process this highly compressed communication that mixes buses, packed lunches, restaurant lunches, cars, and the weather. If the husband addressed this question to anyone other than his wife and expected that the person would comprehend his question, he would be judged to be suffering from a thinking disorder—his thinking would seem disordered to the listener. It is noticeable that what seems like a thinking disorder to the casual listener makes sense if the listener is made aware of the verbal compression the husband and wife have worked out. The odd question makes perfect sense to anyone to whom the situation has been explained. What the husband is really asking in his initial question to his wife is not about packing lunches or taking buses, but about the expected weather conditions that day.

GOOD THINKING—BAD THINKING

When we describe someone as a "great thinker," we pay that person a compliment and judge that his or her thinking is superior to average, both quantitatively and qualitatively. At the other extreme, we apply the term *mindless* to someone who doesn't think things through, who seems not to think at all in the face of a particularly demanding mental challenge, who reacts impulsively without forethought in response to matters that demand the exercise of induction or deduction. The accusation that a person doesn't think things through implies that thinking occupies a timeline—the more difficult the challenge, the more time required for thinking it through. People with attention problems can't think efficiently because they can't maintain sufficient focus over time to consider different possible solutions.

The medieval scholastics and others established over the centuries the discipline of logic, which is as a set of rules for correct thinking. The discipline is based on a set of beliefs that a person can achieve truth if only correct thinking is employed; if the thinking isn't precisely logical, errors occur. But such a belief creates a problem of its own. Since machines can follow some of the rules, machines can be said to think, as narrowly defined. For instance, if we accept rapid computation as an example of thinking, then a supermarket cash register can think more efficiently than a majority of the customers in the store.

The mention a moment ago of thinking occurring along a time frame has its correlate in the spatial dimension as well. As an example of the influence of spatial metaphors on thinking, consider the first of the thinking exercises I promised.

Imagine you have received and read the office email announcing, "Next Wednesday's staff meeting has been moved forward two days." On what day would you appear for the rescheduled meeting?

Your selection of either Monday or Friday is determined by what psychologist Lera Boroditsky terms an *ego-moving perspective* or a *time-moving perspective*. If you think of yourself moving forward in time (the ego-moving perspective), then moving the meeting forward means moving it in a forward-moving direction—from Wednesday to Friday. If you think of time as an impersonal force that is moving toward you (the time-moving perspective), then moving the meeting forward means moving the meeting closer to you from Wednesday to Monday.

Our propensity when thinking to revert to spatial and temporal metaphors illustrates that language is often a determiner of our thinking.

The linguist Benjamin Lee Whorf suggested in the 1930s that language shapes the way we think and speak about the world. The more sophisticated the language, the more nuanced the distinctions. Doctors and lawyers employ commonly encountered words (such as *headache* or *property*) but use them in more subtle ways. A similar situation occurs in adults who have learned to speak a second language. No matter how expert they become in the second language, confounding variables may arise. This is especially true when it comes to idioms—phrases that only make sense in their native language.

A friend of mine, a native French speaker with excellent fluency in English, was dismayed when she heard someone in a group who had applied to several colleges describing a small liberal arts college as a "dark horse." The speaker meant that the college didn't have the cachet of an Ivy League college, but it was a very good one and would be acceptable to her. The speaker's point was lost on my friend simply because she had never encountered a French equivalent to "dark horse."

At this point, I hope I haven't led you to believe that concrete thinking has no place in a normal mental landscape. Actually, that's not true. There are occasions when thinking abstractly leads to erroneous conclusions that could have been avoided by taking a more concrete approach. Consider this challenge drawn to my attention by puzzle master David Book: Suppose you have an encyclopedia consisting of twenty-six volumes, one for each letter of the alphabet. They are arranged in order on a long shelf with the A volume on the left and the Z volume on the extreme right. Which of two entries, "psychiatry" or "psychology," would be closer to the A volume? Since the letter *i* in psychiatry comes before the letter *o* in psychology, the answer would be psychiatry, right? To see why this isn't correct, write the terms *psychiatry* and *psychology* on two separate pieces of paper and insert them in this book, arranged with psychiatry first and psychology second as they would appear in a volume with the subjects alphabetically arranged.

Now close the book and replace it on the shelf. Notice that when doing this, you are turning the book around. The first page of the book is on the right, and the last page is on the left. So now *psychology* is on the left and closer to the A volume, and *psychiatry* is on the right and farthest from the A volume.

Although over your lifetime you have placed hundreds and perhaps thousands of books on shelves, it is likely that you didn't appreciate this transformation until you actually did it. You had to be thinking and acting concretely.

Here is another example of the value of concrete thinking: "Rearrange the letters of NEW DOOR to make one word." Before you do too much rotating and shuffling of the seven letters, just concentrate of what is being asked of you. This second challenge requires not only the use of concrete thinking but also your frontal lobe's ability to suppress those automatized responses that are usually activated in letter-rearrangement puzzles. The answer is ONE WORD, a rearrangement of all the letters in NEW DOOR.

So let's end the chapter with a particularly enlightening example. If you think about the previous examples, you should be able to figure it out. It comes from philosopher Patrick Grim:

> A cowboy rides into town on Tuesday. He stays in the town for exactly 3 days. The first day he works in the general store. The second day, he works at a stable. On the third day he hangs out in the Sheriff's office. He leaves promptly the next day and rides out of town on Tuesday.

That seems strange doesn't it? Three days from Tuesday is Friday, not Tuesday. Yet, the narration is correct as written: He rode in on Tuesday and three days later rode out on Tuesday. The example illustrates the powerful effect of *mental set* or *expectancy*. The narrative is deliberately set up to get you thinking in terms of days. But in the narrative Tuesday isn't a day but the name of the horse that carried the cowboy into town. Once that fact is incorporated, the passage makes perfect sense.

Notice what's required in all of these examples is an ability to shift back and forth from concrete to abstract thinking. If you remain fixated on Tuesday as the name of a day, you are incapable of appreciating it as a name for a horse. I highly recommend working with logical and word puzzles to serve as an antidote against dementia and keeping your thinking supple. (In fact, I wrote a book several years ago aimed at this with puzzle master Scott Kim: *The Playful Brain: The Surprising Signs of How Puzzles Improve Your Mind*.)

Sometimes it's necessary to resort to actually carrying out the process described in the challenge. Here is an old chestnut that will either be immediately obvious to you (probably as a result of having heard it before) or strangely baffling. If you don't immediately come up with the answer, you can arrive at it by performing it as a homespun experiment.

"A drawer contains 23 black socks and 7 brown socks. How many do you have to take out to be 100 percent sure of having at least one pair of the same color?"

Obvious, right? You only need to draw out three socks to have your answer in the form of a color-matched pair.

So effective thinking always requires that we consider words from the point of view of their usual usage (Tuesday is the name for a weekday), along with other alternative uses, no matter how rare (Tuesday as the name of an animal).

Here is one more: "What occurs twice in a moment, once in every minute, and never in a billion years?" Stop reading here. Ponder it for a minute or so. Got it? Here is a hint: re-read the sentence. The key is the occurrence of the letter *m* in each of those words: twice in *moment*, once in *minute*, and never in a *billion years*.

Each of the puzzles and conundrums you just worked on have drawn from the powers of your frontal lobe—the thinking lobe. So, if you want to strengthen the frontal lobe, work on puzzles like these. Puzzles (crossword, jigsaw, sudoku, wordle, etc.) provide you with the best chance of keeping your frontal lobes functioning normally. Remember when I mentioned earlier that I have never encountered normal memory functioning in a person with dementia? The same holds true for frontal-lobe functioning. If the major frontal lobe functions are kept highly tuned as a result of repetitive practice, this will lessen the odds of dementia.

How humbling it is to realize that one's most advanced and evolved mental processing depends on the normal functioning of two specific parts of the brain. When either the frontal or temporal lobes—or both of them at the same time—malfunction, Alzheimer's and the other dementias may follow. Keep these frontal and temporal functions in mind as we next explore what's going on in the other major dementias.

CHAPTER VI

DEMENTIA AND ITS VARIATIONS

The phrasing "Alzheimer's and other dementias" is frequently used throughout this book. Since none of these dementias occur as frequently as Alzheimer's, why even discuss them in a book focused on ways of understanding, preventing, and delaying Alzheimer's? For one thing, each variety of dementia shares common elements with Alzheimer's, along with other features relatively specific to that dementia. For instance, memory is frequently affected by Alzheimer's. But memory loss is *not specific* to Alzheimer's; memory failures occur in all the dementias at some time in the course of the disease. Some dementias are primarily *neuropsychiatric* and affect behavior and emotions. Others are characterized by what neurologists refer to as *motor disturbances*: weakness, tremor, clumsiness, falls, or extremely slow movement. But eventually all of dementias contain elements of these two varieties of neuropsychiatric and motor.

Consequently, recognition of one dementia from another requires the ability to distinguish subtle differences. This is important because specific treatments may be required for each. Time of onset and life expectancy also differ among dementia types. Another distinguishing feature is the natural history of a particular dementia and the affected person's response to the disease.

In all cases of dementia, something a person says or does eventually brings his or her condition to the attention of others. You don't have to be a neurologist or psychiatrist to appreciate at some point in the illness that something is wrong. What follows are the four most frequently encountered dementias after Alzheimer's, along with ways of distinguishing them.

THE FINAL DAYS OF ROBIN WILLIAMS

In October 2013, ten months before his death, the comedian and actor Robin Williams complained to his wife, Susan, of a "gut discomfort" that made him anxious and fearful. He was unable to pin down what was bothering him. Over the next several months, he experienced panic attacks and, most disabling for one in his profession, an inability to memorize his lines for a role he was preparing for as Teddy Roosevelt. In retrospect this seems especially ominous, since only three years earlier Williams had effortlessly memorized without hesitation or error page after page of lines for his role on Broadway in *Bengal Tiger at the Baghdad Zoo*.

One moment Williams was capable of discussing and reasoning about the best course of action to follow in order to find the cause of his difficulties. "Five minutes later he was blank, lost in confusion," wrote Susan Schneider Williams.

Other symptoms occurred helter skelter: paranoia, insomnia, and loss of the ability to smell. Most distressing was the depression, which only grew worse with each passing day.

For over a year, Williams underwent exams and blood tests and scans, tried medications, worked at physical therapy, employed a personal trainer, even tried yoga and self-hypnosis, all to no avail. The underlying illness continued to progress.

At this point, doctors reached a diagnosis of Parkinson's disease. His physical symptoms (a tremor, a stooped gait with forward flexion and decreased arm movements) occur in both Parkinson's and the disease Williams was eventually diagnosed with.

On August 11, 2014, Williams was found dead in his home in Paradise Cay, California. The autopsy report established the immediate cause of death as suicide by asphyxia (loss of oxygen) due to hanging. Examination of the brain revealed the underlying illness: Lewy body dementia (LBD).

LBD, the second- or third-most common dementia after Alzheimer's, is the most difficult dementia to recognize. One has only to consider that Robin Williams, a wealthy, world-recognized performer without financial constraints that would limit his search for solutions, died before the correct diagnosis was reached.

As with Robin Williams, people with LBD exhibit three of the following core features:

- Fluctuations in the clarity of their thinking with pronounced variations in attention and alertness (Williams's difficulty memorizing scripts; being "lost in confusion," in his wife's words).
- Sleep disturbances.
- One or more features of Parkinson's disease, such as slowness of movement, tremor, or rigidity. (Williams displayed all of these, hence the mistaken diagnosis of Parkinson's disease.)
- Recurrent, well-formed, and detailed hallucinations. ("A year after he left, while I was speaking with one of his doctors who reviewed his records, it became evident that most likely he did have hallucinations, but was keeping that to himself," wrote Schneider Williams.)

It is likely that Williams met all four of these criteria. He also displayed all the "supportive features used to establish the diagnosis," such as apathy, anxiety, depression, delusions, loss of smell, constipation (a sign of involvement of the autonomic nervous system), and transient episodes of unresponsiveness.

Complicating the diagnosis of LBD in Robin Williams was his documented history of cocaine and alcohol abuse. Although Williams had been treated within a year or two of his death for alcoholism, there is no indication that after giving up cocaine he ever returned to it. But even among the people with no history of drug or alcohol abuse, the diagnosis arrived at in the case of LBD is frequently a psychiatric one. (Williams also met many of the criteria for bipolar disorder.)

Of all the indications of LBD, hallucinations stand out. As with Williams, many people afflicted with hallucinations don't talk about them, lest they immediately identify themselves as "crazy" to others. The hallucinations can vary from detecting something or someone passing fleetingly in their peripheral vision; seeing small fuzzy animals, such as a deceased pet, for instance; mistaking a coat on a hanger for an actual person; to discerning images emerging from wallpaper. As the disease progresses, the images increase in frequency: little people ("midgets" as one patient referred to them), children, or cuddly animals.

Parkinsonian features occur in over 50 percent of people with LBD, including a generalized slowing of movement with little or no swinging of the arms as occurs in a normal gait, body rigidity, tremors, and frequent falls.

As with Robin Williams, doctors can only make an absolutely certain diagnosis at autopsy. Pathological examination reveals generalized brain atrophy, predominantly affecting the emotional circuits (limbic system), along with the presence of Lewy bodies—an abnormal aggregation of breakdown products within the neuron. To date, more than seventy molecules have been identified in Lewy bodies, with a compound named alpha-synuclein as the major constituent. Since Lewy bodies are also present in Parkinson's disease, it's no surprise that Lewy body disease is often mistaken for Parkinson's disease.

Although not as conclusive as an autopsy report or brain biopsy, certain biomarkers are highly suggestive of the disease. Testing for one of them is easy to administer. An overnight EEG study may explain one of the puzzling behaviors almost universally encountered in LBD: rapid eye movement sleep behavior disorder (REMSBD). Suddenly the person afflicted with REMSBD may leap out of bed during the night and begin running until abruptly stopped by a wall or other barrier. This happens because the temporary paralysis of limbs that occurs during normal sleep does not occur in the person afflicted with LBD. As a result, a frightening dream may lead to actual escape maneuvers, made possible in the absence of the normal sleep-induced paralysis.

LBD is also suggested by a relative preservation of the medial temporal lobe structures on imaging studies.

If the diagnosis is established, symptomatic care can improve things to some extent. But sadly, no cure exists.

From the time of diagnosis until death, the lifespan of the person with LBD is greatly foreshortened compared to a person with Alzheimer's disease.

WHO WON AND WHAT WAS THE SCORE?

Frontotemporal dementia (FTD) is another of the most common neurodegenerative dementias after Alzheimer's disease and dementia with Lewy bodies (DLB). It encompasses 15 percent of dementias and is estimated to affect 50,000 to 150,000 Americans. Typically, it has an early onset, when people are in their forties to sixties, and in fact is the most common early-onset form of dementia. Cases have been reported with onset as early as in the twenties. The progression is sufficiently slow that about 3.6 years pass before an accurate diagnosis is made. But the most striking aspect of frontotemporal dementia (FTD) are its initial signs: behavioral changes that almost invariably lead to an initial misdiagnosis of depression or another mental health condition. Unfortunately, no effective treatment exists, and the disease steadily progresses.

As the name indicates, frontotemporal dementia (FTD) affects primarily the frontal and the temporal lobes. Judgment, abstraction, thinking, and behavioral regulation are the most important frontal-lobe functions.

In the very early stages, the behavioral manifestations aren't that easily distinguishable from normal responses. For instance, if the amygdala and other parts of the limbic system (a circuit devoted to the internal appreciation and outward expression of negative emotions) activate inappropriately, any one of us may be given to temper outbursts and verbal or even physical aggression. Such an outburst is prevented in the healthy person by the *damping influence* exerted by the frontal lobes: We talk ourselves down, much as a friend or handler might. "Relax. Your reaction to such a trivial event is excessive." But in FTD, the frontal lobes are damaged and incapable of exerting the damping function.

It is not uncommon for temper tantrums and outbursts to be the first indication of frontotemporal dementia.

One of my own patients with an eventual diagnosis of FTD with predominantly frontal lobe involvement was referred to me after a dinner-party kerfuffle. He had not wanted to attend the dinner and expressed a strong desire to stay home to watch a basketball game on television. His wife made the perfectly reasonable suggestion that he record the game, so that he could watch it later that night after they returned home. He agreed but added that he didn't want to know how the game ended until arriving home when he could watch it on the recording.

Things went well until one of the guests excused himself from the dinner table and, while returning from the bathroom, glanced at the television set at the moment of the final play of the game. When the guest returned to the table, he sat down at his place and said, "If there are any basketball fans here, this is how it ended" and then gave the final score. My patient erupted—shouting, rising, from his chair, and even pushing the man.

More similar episodes occurred over the next several months—culminating in the FTD diagnosis. Added to the temper tantrums were symptoms of memory failure, disorientation, literal concrete thinking, blunting, lack of drive/motivation, loss of the ability to perceive social cues (either from other people or situations), and finally loss of the capacity for imaginative identification with others (foreshadowed by that emotional outburst at the dinner party).

In FTD three variants exist. My patient suffered from the *behavioral variant*: psychiatric symptoms of thought and behavior stand out. *First* to appear, as with my patient, is behavioral disinhibition. He was not able to put the basketball game into perspective and struck out with greater intensity than one would expect to see even in a coach or a relative of one of the players. Over the next few months, irascibility was replaced by the *second criterion*: apathy or inertia. Often confused with depression in the minds of even seasoned psychiatrists or neurologists, an apathetic person simply cannot summon any interest or enthusiasm—everything seems blah. The sign for apathy is hands held in a palm-up position, accompanied by a "so what?" shrug of the shoulders.

The final criterion is loss of sympathy or empathy for others. We interviewed one man with frontotemporal dementia several years ago while filming *The Brain Series* for public broadcasting. He had this to say while his wife sat beside him on the bed: "She's got breast cancer. So I don't know how much longer I can depend on her for help," spoken in an emotionless, almost robotic tone of voice. No indication of concern or words of comfort was present, just a short, blunt assessment that, unsurprisingly, led to his wife's tearfulness.

"Frontotemporal dementia is unique," according to Brad Dickerson of the Massachusetts General Hospital FTD Unit & Center for Translational Brain Mapping. "It is more than a loss of cognitive abilities. It includes changes in personality that can be profound. The affected individual may find it hard to navigate, display an altered sense of humor, lack compassion, demonstrate irritability and have failings in social etiquette, which lead to a severe communication barrier."

BRUCE WILLIS'S SEARCH FOR WORDS

If the temporal lobe is involved, especially the left temporal lobe, speech and language are affected. Aphasia—difficulty understanding language, whether spoken or heard, is the first symptom. This might be followed by behavioral symptoms.

In March 2022, internationally recognized film star Bruce Willis retired from acting because of difficulty speaking, which was diagnosed as aphasia. When you think about it, what could be more incapacitating for an actor than aphasia? One of the requirements for the profession is an ability to come up with the right words delivered lightning fast, razor sharp, and with nearly perfect articulation. Gradually, Willis lost this ability and, despite all the prompts and gimmicks available to help challenged actors, he eventually fell below an even minimally acceptable verbal performance.

As a consequence of aphasia, whether in a thespian or a member of the general public, the affected person experiences difficulty constructing grammatically and syntactically correct sentences, finding the correct words, and understanding the meaning of words. In response, the person with these temporal-lobe difficulties tends to avoid conversations and, except among accepting and patient listeners, often avoids people altogether. Primary progressive aphasia (PPA) is the term for difficulty expressing or understanding language. Over the next eleven months, the difficulty increased to the point neurologists diagnosed Willis with PPA, secondary to frontotemporal dementia.

Unfortunately, frontotemporal dementia cannot be cured; only the florid behavioral symptoms, such as outbursts of aggression traceable to frontal-lobe dysfunction, can be treated. Although the cause is unknown, frontotemporal dementia (FTD) may be related to two abnormal proteins, known as tau or TDP-43, accumulating in the brain. So far at least fifteen gene mutations have been discovered to be associated with the illness. This raises the possibility that different expressions of FTD are the result of different causes.

Alzheimer's and frontotemporal dementia (FTD), along with Lewy body disease, are all in the category of neurodegenerative brain diseases. All involve abnormal behavior resulting from deterioration in the structure, chemistry, or electrical functioning in the brain. Most start with behavioral-psychiatric symptoms.

"Rather than lacking scientific relevance, psychiatric symptoms are key to the understanding of dementia and are reflective of the anatomy and chemistry of the disease," according to Bruce L. Miller, MD, director, Memory and Aging Center at the University of California San Francisco Weill Institute for Neurosciences.

"WHAT'S GOOD FOR THE HEART IS GOOD FOR THE BRAIN"

The heart and the brain are intimate partners working together to maintain healthy cognitive function. Even though almost all of our attention up to this point has been focused on the brain, normal brain functioning is almost impossible if the heart and its extensions—the cardiovascular system—are diseased. Not only can **vascular cognitive impairment** lead to its own form of dementia (vascular cognitive impairment and dementia, or VCID), but it also plays at least a minor role in all of the other dementias.

The first variety of VCID is stroke. The end result of stroke, if it is not successfully treated, is a destruction of tissue (infarction) at one or more sites of the brain. If it affects part of the brain responsible for movement of the arm and leg on one side, for instance, the person is paralyzed on the opposite side of the body. If the stroke involves the very back of the brain where the visual fibers come together, the result is blindness, which on rare occasions may be accompanied by denial of that blindness (so-called Anton syndrome; more on that later).

The second variety of VCID demonstrates a slow rate of progressive impairment that on imaging shows evidence of vascular brain injury. At autopsy, VCID shows few or no plaques or tangles, but multiple areas of destruction (lacunae) along with large infarcts over the cortex and smaller infarcts in the areas beneath the cortex (subcortical areas).

Pure Alzheimer's and pure VCID are hard to distinguish from each other since they share common disabilities. Both are marked by a decline in previous functions sufficient to interfere with social relations, as well as loss of the ability to carry on everyday activities. Often, a CAT scan or MRI scan distinguishes the two diseases. While an Alzheimer's brain image may show little other than some mild-to-moderate generalized atrophy (loss of brain tissue), the vascular impairment of VCID typically shows extensive arteriosclerotic disease, which alone or in combination with Alzheimer's forms the basis of a resultant dementia.

Meet Harold, a man is his late seventies. Over a five-year period, Harold developed memory problems; language difficulties, especially understanding other people; dragging of his left foot secondary to a stroke; chronic problems with irritability and even anger; and depression. The key diagnostic component here was the stroke. None of the other problems was definitive enough to suggest anything other than the onset of a dementia of undetermined type. At autopsy five years later, Harold's brain showed moderate-to-severe atherosclerosis, along with scattered infarctions. Also found were the tangles and senile plaques of Alzheimer's. The final diagnosis? According to the autopsy, "The preponderance of findings related to VCID with relatively moderate evidence of Alzheimer's."

As Harrold's case illustrates, VCID can combine with other dementias, in this case Alzheimer's.

VCID is a bit of a paradox. Although it is one of the most frequent causes of dementia, perhaps the second-most frequent, it often slips under the radar because in many instances it is not dramatic in onset, usually showing few neuropsychiatric indicators. Harold's was a typical case showing irritability and depression but no delusions or hallucinations that would be guaranteed to capture the attention of relatives, friends, and doctors alike. In such cases, VCID is characterized by the onset of a grindingly slow period of decline secondary to small blood vessel infarction.

VCID is of particular interest in another way. It is the only form of dementia that over the past twenty-five years has shown a universally recognized *decrease* in frequency. "We did find substantial decreases over time in brain atherosclerosis and arteriosclerosis," wrote Francine Grodstein, of Rush University Medical Center, in a 2022 issue of *JAMA Neurology*. Since both atherosclerosis and arteriosclerosis are the villains in cardiovascular disease, the findings suggest the efficacy of the national efforts over the past several decades to reduce vascular risk factors and improve vascular health. This is an example of the oft-cited mantra "What's good for the heart is good for the brain."

Of even greater importance is what these findings mean for dementia control in general. It *is* possible to reduce at least one form of dementia (VCID) by following a lifestyle program (diet, exercise, diabetic control, etc.) that comprises the components of the heart diet recommended by the American Heart Association. Further, when we consider that VCID is often a contributor to the other dementias—especially Alzheimer's—we thus have strong support for believing that lifestyle can modify, delay, or perhaps even prevent at least one form of dementia and possibly more.

So far, we have described the four most common causes of dementia. Alzheimer's disease is in first place by a wide margin, followed by Lewy body dementia, vascular dementia, FTD, and Parkinson's dementia. Despite important differences, what do all these have in common? Age, of course. All of them involve older people (with the exception of rare cases of inherited early-onset Alzheimer's and many cases of FTD). With these exceptions, all the other most commonly encountered forms of dementia that we have discussed so far occur beyond middle age.

So are younger people immune from dementia? A decade or so ago, the answer would have been "Dementia in the young (teens to forties) is very rare." That answer is not correct anymore.

Thanks to widespread participation in football, soccer, hockey, and even boxing, some athletes participate in sports for two decades or more. During their careers, even amateur athletes may suffer repeated blows to the head, resulting in chronic traumatic encephalopathy (CTE).

GETTING YOUR BELL RUNG

A chain of repetitive head injuries sets off in the brain a cascade of changes resulting in the accumulation of an abnormal form of tau protein. This waste product accumulates within the brain cells and spreads even after the head injuries stop. Since CTE has a unique pathology along with a preferential location in the superficial layers of the cortex, CTE can only be confidently diagnosed by an examination of the brain after death. A *specific abnormal variation of tau* is only seen in the brains of people with a history of repetitive brain trauma. In contrast to Alzheimer's, the CTE brain might only feature amyloid plaques. A microscopic examination is useful for distinguishing the two diseases.

Here is the sequence: Head trauma leads to an injury of the small blood vessels (arterials) within the meninges (outer layering of the brain). The tau accumulates in the neurons surrounding the arterials. Next, leakage of protein takes place, leading to inflammation, which in turn leads to additional tau accumulation. Eventually, due to the repetitive head injuries, the brain's natural clearance system for the elimination of waste products like tau is overwhelmed.

Since differences of opinion exist about how many head injuries are sufficient to cause CTE, researchers measure, as an alternative, the number of years an athlete participates in a particular sport. The result is sobering.

The odds of CTE double with every 2.6 years of competitive full-contact football. In regard to soccer, an increasingly popular sport in the United States, dementias were 62 percent more common among soccer players than in people who never played soccer. Even more worrying, the threat was not evenly distributed among players. Goal keepers didn't show any increase in risk, but outfield players did—more than a hint that heading the ball may underlie the later development of dementia.

According to a study carried out at the Karolinska Institute of Stockholm and published in *Lancet Public Health*, the greatest risk of dementia occurred among outfield soccer players with long careers. Based on the health records of 6,007 soccer players from the top divisions during the period extending from 1924 through 2019, 8.9 percent of the elite players developed dementia compared to 6.2 percent of nonplayers (citizens drawn from the general population). The founder of the study properly noted that his findings did not establish a cause-effect relationship: "Even if we would have perfect data on causality, what to do with it is a matter of values and a decision for the . . . individual players to make."

The sequence of the disorders associated with CTE also differ remarkably from Alzheimer's, which usually, but not always, begins with memory issues. In CTE the initial changes involve emotional and behavioral control, agitation, and a "short fuse"— outbursts of impatience and anger associated with verbal and in some cases physical abuse. Formally identified as "neurobehavioral dysregulation," such symptoms both mark the start and determine the probable course of the illness. Interwoven with these emotional-behavioral changes are difficulties in short-term memory, including difficulty forming new memories; rapid forgetting; and what's known as *confabulation*, the introduction of memories for events that never happened. These cognitive changes usually occur a few years after the emotional outbursts. If this sequence and timing aren't kept in mind, the mistaken diagnosis is almost always Alzheimer's disease. The final stage of CTE consists of dementia ranging from mild to severe.

Public responses to the widespread information about CTE has already led some parents to discourage their sons and daughters from participating in football or soccer. Is this an overreaction? Unfortunately, a definitive answer to that question is impossible. As already mentioned, not every player sustains the same amount and degree of head trauma. This threat, as mentioned, varies with the sport, the position played, and the frequency and ferocity of body contact. These factors can't be measured simply by equating the number of games played with time on the field and frequency of identified concussions. Practice sessions along with regular games should be included, but they frequently aren't.

Since a single head injury can be the underlying cause of CTE, what about people suffering a mild minor concussion in a motor vehicle accident? Are they also susceptible to CTE? This issue is a matter of active litigation, usually taking the form in court of cross-examination questions like "Isn't it true that the head injury sustained by my client in the motor vehicle accident increases the odds of him later developing dementia?"—a hard question to answer definitively, based on what we know and what we don't know today about head trauma and dementia.

Since the development of sophisticated brain imaging, neuroscientists can prove the brain can be injured by trauma. I'm not referring to major trauma to the head, often resulting in skull fractures or prolonged unconsciousness. You don't have to be a doctor to suspect such severe head trauma is likely to lead to some degree of brain damage. I'm referring here to the milder kinds of head trauma, commonly described as the "dings," "seeing stars," or "getting my bell rung."

These everyday terms refer to the experience of concussion. The word *concussion* comes from the Latin *concutere*, meaning "to shake mildly." Although the term is in common usage, understanding of its meaning is not. So before probing any deeper into CTE, it's helpful to have some understanding of a concussion.

Repetitive concussions occur most frequently in competitive sports. Since sports are largely in the province of the young, CTE represents the major form of dementia encountered in the young (a time range from the teens to the late thirties and early forties).

Research on concussions over the past decade demonstrates that loss of consciousness is not even necessary (less than 10 percent of concussions involve a loss of consciousness). This so-called subconcussive trauma to the head, if repeated often enough, can lead to permanent brain injury short of dementia.

"Growing evidence suggests that even after one season [for the athlete], repetitive sub-concussive trauma can lead to cognitive, physiological, metabolic and structural changes," according to Robert Stern, PhD, of Boston University Alzheimer's Disease Research Center. Translation: if you incur a sufficient number of subconcussive injuries, your brain will undergo changes, leading to a deterioration in the quality of your thinking, along with its chemistry and structural integrity, culminating in CTE.

Concussion is associated with specific processes resulting from the physical impacts to the head. Helmets, incidentally, do a great job of preventing skull fractures, but do not protect the brain from concussion. Why?

Contrary to common belief, the brain is not tightly tethered to the skull. It is more loosely suspended within the cerebral spinal fluid, which serves as a buffer against the effects of mild trauma (like hitting one's head on a cabinet in the kitchen). But under conditions of more severe trauma, such as experiencing a football tackle, the brain is susceptible to linear, lateral, and rotational forces within the skull, which produce rapid brain acceleration and deceleration leading to a direct impact against the inner surface of the skull. These sudden changes in velocity and direction also lead to a shearing (tearing) of the cells of the brain and their extensions (tracts), which, in turn, contribute to changes in ionic balance (sodium, potassium, and other ions) and metabolism (release of chemicals that slow the speed of the nerve impulse). This combination of factors results in the mental sluggishness of the concussed—routine questions ("Where are you right now?") may be answered correctly, but the answers are preceded by a pause lasting several seconds. Helmets play no role in mitigating any of this.

In the majority of cases, concussion is followed by complete recovery, but the prognosis for those sustaining repetitive concussions is less favorable. The earlier in life an athlete begins his career in a sport like football, the more head injuries and concussions he will experience, and the more likely it will be that he will develop chronic traumatic encephalopathy.

CTE has a long history, dating at least to the early twentieth century, when it was observed principally in boxers. Studies of boxers dating from late 1920s described the punch-drunk fighter, suffering from what was once named "dementia pugilistica," now included under the diagnosis of CTE. It is believed that the repetitive brain trauma activated a cascade of chemical changes and structural degeneration of brain tissue.

Starting from my early interest in boxing at age twelve or so and continuing to the present day, I have encountered along the way several professional boxers with the thick slurred speech and slaphappy dispositions that often accompany CTE. I have even treated as a regular patient a former heavyweight champion. Perhaps it strikes you as odd, perhaps even hypocritical, that someone like me, trained in brain science, should maintain an interest in something universally recognized as destructive to healthy brain functioning.

The short answer is that my lifelong interest in boxing started as a response to bullying. I grew up in the era marked by a "Let the kids work it out between themselves" attitude toward childhood bullying. I had to learn to defend myself. After some boxing lessons and a triumphant fistfight in the school yard, which I remember with cinematic clarity to this day, the bullying stopped. As an unanticipated and probably unfortunate side effect of learning some of the techniques of the "sweet science," I've maintained an interest in the sport all my life. Does that make sense? You don't think so? Perhaps you're right. But as I will discuss in the next portion of this book, even if we are reasoning creatures, we are not exclusively or even predominantly always logical creatures.

CHAPTER VII

LOGIC, STATISTICS, AND DEMENTIA

Now that I have provided the history of dementia and its variations, it's time to take up the most intriguing question of all:

> Does dementia represent a condition completely at odds with normal functioning (a qualitative difference), or does dementia represent the extreme of a continuum starting with perfectly normal cognition and extending to a severe mental impairment (quantitative difference)?

Notice these questions aren't wholly scientific ones. When we search for causes, we enter the realm of logic. You don't have to be formally trained in logic (although it helps) to wind your way through the bramble bushes that interfere with logical thinking.

To come up with conclusions about Alzheimer's, it is useful to examine the thought processes we employ in our research. Mistaken assumptions tend to lead to mistaken conclusions.

If we don't think logically, we won't come up with the cause, cure, or palliative treatment for Alzheimer's, nor will we be able to estimate our chances of coming down with it.

A FALL DOWN THE STAIRS

What is perhaps the greatest challenge to logical thinking? Distinguishing a *cause* from a much more commonly encountered *correlation*. We carry umbrellas on rainy days, or even on days when rain is considered only a possibility. But rain and umbrellas involve correlations and not causation. Some people are less discombobulated by the threat of rain, will accept the risk of rain on a cloudy day, and leave home without an umbrella. After all, if it rains, one can always buy a cheap umbrella from a street vendor.

This course of action is neither more nor less logical than the person who at the mere mention of a possibility of rain carries an umbrella to work. Such distinctions are not really about logic but individual attitudes towards risk-taking, or in this case, inconvenience-taking.

Let's move on to situations when a cause can be distinguished from a correlation.

Causation always involves the explanatory, providing an answer to the simple question "Why did this happen?" You slip on a stair and fall on your back, sliding down several stairs. The CT scan shows two small fractures—a lot of pain but no surgery required. The relationship between the vertebral trauma and the fractures is causative—no fall, no fractures.

Correlation or causation is not always so easily distinguished. "Does alcoholism cause smoking?" The two are frequently encountered together, but there is no evidence of causation. Both smoking and drinking involve repetitive, harmful, and habit-forming behaviors that can easily spin out of control. So it should not be surprising to learn that they frequently occur together. But one does not *cause* the other. The relationship is strictly correlational.

Correlations can be positive when both variables move in the same direction. "The more calories you consume, the more weight you will gain" (forget about exercise and general activity level for the moment). Correlations can also be negative, with both variables headed in opposite directions. "The more money you spend without replenishing it by depositing more, the lower your bank account." While correlation doesn't imply causation, causation always implies correlation. The trick is teasing out the difference.

Correlational links are like circumstantial evidence in a crime investigation. Circumstantial evidence, a form of correlation, can be persuasive but not 100 percent convincing. You need that smoking gun.

If someone comes down with Alzheimer's, is this a predictable event that can be explained based on causation, or is it due to the influence of an unknown number of factors? In a way, coming down with Alzheimer's can be compared with being struck by lightning.

ONE IN A MILLION

On August 4, 2022, four people were struck by lightning during a massive thunderstorm while they huddled under a tree across from the White House. Six lightning bolts within half a second killed three. The lone survivor, 28-year-old Amber Escudero-Kontostathis, was waiting prior to the lightning strike for her husband to pick her up for her birthday celebration.

While the inadvisability of standing under a tree during a lightning storm is well known, the knowledge of the exact consequences of ignoring that rule varies from person to person. "I always thought, like, if a tree were hit by a lightning, it would catch fire and you run from the fire," said Escudero-Kontostathis in a television interview after the incident.

In addition to the variations in knowledge about lightning from one person to another, several other factors interfere with logical thinking. First, people often seek shelter under trees during electrical storms and they usually aren't struck by lightning. "I'll probably be able to get away with standing under this tree for a few minutes," they may rationalize. In the overwhelming majority of cases, they will get away with it.

Further, different correlations about the likelihood of being struck by lightning differ among people based on their limited experiences. Thanks to the *familiarity effect*, anyone in the near proximity of Escudero-Kontostathis's lightning injury will in the future exaggerate in their minds the odds of being struck by lightning when caught out in a rainstorm (even in a storm in which the occurrence of lightning is not observed). These same people may search databases such as LightningMaps.org, which provides a region-by-region map with updates every twenty minutes of lightning strikes anywhere in the United States. They may read and even commit to memory such facts as Florida's reputation as the lightning capital of the United States, which over the last fifty years has hosted two thousand incidents of humans struck by lightning, with thirty deaths. It might even sway their decision to not visit Florida or, if already a resident, to move out of the state.

But what about people who only heard about the White House lightning strike by reading about it in the newspaper or encountering a description of it on television? I would guess they would be much less likely to consult LightningMaps.org (I didn't until researching this). They are also more likely to be reassured by the statistics such as this: between 2003 and 2015, only about thirty-five deaths per year by lightning strike occurred in the United States. It is even more likely they will be reassured by the commonly ballyhooed statistic that the overall odds of being struck by lightning are about one in a million.

So given all this, what is the likelihood of *you* being struck by lightning? This is a far more difficult question to answer than it at first appears. For one thing, the odds in individual cases are far harder to estimate in terms of statistics alone, without any individual qualifiers. The statistical odds of an undetermined person within the general population being struck by lightning is far different from the odds that a specific person (you or me) becoming lightning-strike victims. We are living, breathing, decision-making creatures who can take specific measures to reduce the odds (more about that in a moment). In contrast, the statistical person only exists on a spreadsheet or other model.

For one thing, the odds are shifted by certain patterns and traits such as never going out in the rain. In addition, the odds vary greatly depending on where you are living. If you live in Montana, the odds of a lightning strike is approximately 1 in 249,550—a good bit higher than "one in a million."

Several factors related to lightning strikes remain unexplained. Males account for close to 80 percent of all lightning fatalities. Attempts to explain this disparity are rife with generalizations. "Because of their behavior, males are at a higher risk of being struck and consequently are struck and killed by lightning more often than females. . . . Males are unwilling to be inconvenienced by the threat of lightning," according to a 2020 statement by the National Lightning Safety Council.

Finally, we come to the outdoor-indoor frequencies. From the commonsense point of view, it would seem hard to be struck by lightning if you spend most of your life indoors. This presumably would hold true wherever in the country you may be living. Actually, one-third of lightning strikes occur indoors. Lightning enters a house by direct strike—through wires or pipes that extend outside the building or by striking the ground near the house. Once within the building, lightning can travel through electrical, phone, radio, or television-reception systems, as well as the plumbing. As a result, people have been killed or injured by lightning while using computers or TVs or simply by touching electric cords. Since the plumbing serves as a conductor, people have died while washing dishes or taking a shower.

Just as you can lower the odds of a lightning strike, you can do the same with the development of dementia. But whatever you do, you cannot *guarantee* not to be affected by either of these.

I've discussed lightning strikes in such detail because the topic provides many parallels with the odds of coming down with dementia. In both instances, the likelihood depends on a concurrence of multiple variables over time, none of which are entirely predictable or controllable.

If this discussion strikes you as a bit theoretical, trust me. It has critical relevance to the thought processes in which we must engage in order to understand the factors responsible for Alzheimer's disease and other dementias.

As another example of a tendency to confuse correlation with causation, consider this: Some people—myself included—find that they perform better on standardized tests if they are permitted to sip coffee or tea while taking the test. So can we conclude that these drinks increase intelligence?

Not really. What they do is contribute to our alertness, focus, concentration, and energy. After the coffee or tea effect wears off, we revert to a less focused mental state. All of this is correlational and not causative. On occasion, the correlation can work in the opposite direction: too many caffeine-containing drinks can induce the jitters or restlessness—just the opposite of the focused, concentrated state we were aiming for. There are other reasons for not settling on one factor as exerting a causative effect. How much sleep did you get the night before? How much preparation have you put forth for the exam? If you are prepared and really know your stuff, your ordinary level of alertness and concentration should do just fine and not need any cognitive stimulants.

Two impediments prevent a correlation from rising to the level of causation. First is the presence of a *confounding variable*—something not measured and often not even considered when deciding whether a linkage is correlational or causative. The sale of ice cream, sodas, and beer correlates with the outside temperature. As a result, the sale of all of them goes up in the summer months. But this link isn't causal, because temperature increases lead to many other heat-related responses: the greater use of air-conditioning, more visits to the swimming pool, cooling off in a shower, etc. All these are correlational efforts related to the rising temperature. Ice cream sales were only one of many correlational factors related to the rising temperature.

Second, sometimes both variables correlate consistently and might actually form a causal relationship, but it's impossible to decide which variable causes a change in the other. This is the so-called *directional problem*. Or the causal question can be narrowed or widened so that totally different influences come into play. What caused you to read this book, for instance? Are you reading it because you are deeply interested in ways to prevent or forestall dementia and Alzheimer's in particular? Or are you reading it because you suddenly had a desire to read a book and for some reason this book was the only one at hand? These very different correlational questions invoke different responses.

In some instances, one explanation may stand out among several others as the "cause." Take the example I mentioned a moment ago of a fall on the stairs. Although we have already decided that a causal link exists between the trauma and the fractures of the bones in the back, the reason for the fall doesn't lend itself to a causative explanation. Here is a graphic of just some of the factors that correlate with falls at home.

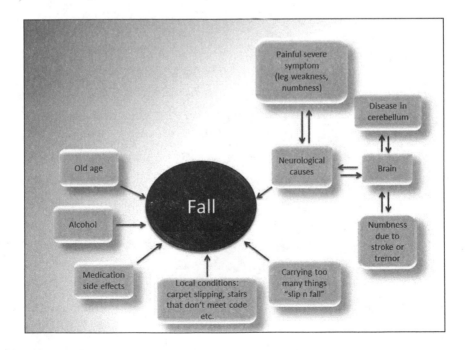

Nor do these separate factors operate in isolation. The injured person may be elderly, have a nervous-system disorder such as diabetes-induced numbness of the legs and feet, and take one or more medications that affect balance. Add to all this a dollop or two of alcohol, and you'll have all the necessary ingredients for a severe tumble down the stairs. With all of these factors in play, there really is no one *cause* of the fall, merely one or more con-tributors. So the factors leading up to the fall are correlational, while the fractures in the back are due to the direct effect of the fall (causative).

Too often we find ourselves attributing causation to some-thing that happened, either good or bad. *Post hoc ergo propter hoc*— after this therefore because of this—is the most common logical error. Because of the structure and functions of our brains, we are explanation-demanding creatures who cannot tolerate uncertainty. Even a false explanation seems somehow better than no explanation at all. We are reasoning creatures but, as with my continued interest in boxing mentioned a few pages ago, not always logical creatures.

THE NOT-SO-MAGIC 99.9 PERCENT

Would you buy a dress or a suit from a local store that 99.9 per-cent of previous purchasers rated five on scale of zero to five (five equaling highest satisfaction)? Would you eat at a local family-owned restaurant that 99.9 percent of the previous customers rated as highly satisfactorily?

If you answered yes to these two questions, you are probably making a good decision. True, the number of dresses and suits sold by the local store, as well as the numbers of meals served at the local restaurant, is small. But if the overwhelming majority of people were satisfied with their clothes and their meals, the like-lihood is high that you will be satisfied as well.

Would you undergo an operation at a hospital with a 99.9 percent safety record? Would you fly on an airplane knowing that the quality and safety rate of commercial airlines was about 99.9 percent?

With the hospitalization and airplane trip, some very different factors are called into play. As the stakes get higher, the likelihood of a desired response must be as close to 100 percent as possible.

To be more specific, if a hospital system on the whole maintains a 99.9 percent quality rating over one year, 144 mistaken medical procedures will occur. Eighteen newborns will be matched with the wrong parents each day; 567 pacemaker operations will be performed incorrectly in a year. A 99.9 percent quality performance in the airline industry would lead to even more disastrous consequences, with over eight hundred commercial airline crashes *every month*. Now let's put this into the context of Alzheimer's disease.

The current accuracy rate of a much-anticipated Alzheimer's blood test is currently 85 percent—quite a difference from 99.9 percent. Yet these tests are expected to be widely available within the next few years. While you are considering the implications of that, please ponder this:

How close should you get if 99.9 percent isn't good enough? The answer involves what's called the *five nines*. No less than 99.999 percent is the quality control demanded for high-risk situations. And the final two nines in 99.9<u>99</u> percent are crucial.

A 240-minute power outage over the space of a year represents an 99.9<u>54</u> percent protection rate against power outages. That's probably acceptable in some cases. In fact, if you worked in an office heavily dependent on computers, I'll bet you have a down-time over 240 minutes a year without experiencing any horrendous consequences. But if you work in a major financing company and the outage occurs at a crucial time, 240 minutes could translate into thousands or even millions of dollars at risk for every minute the stock exchanges are inaccessible. Within the space of 240 minutes, depending on how the outages are spread out, we can be talking of a loss of millions of dollars. So, 99.954 percent just isn't a good enough reliability rating. Less than five minutes and fifteen seconds of downtime a year equals 99.999 percent. Even then, the consequences could be severe, if that downtime occurred at the wrong time.

FAULTY REASONING

Equally important is the faulty reasoning that often is applied to statistics. Consider this example of faulty reasoning published in *The Lancet*, first detected by Bjorn Lomborg, president of the Copenhagen Consensus Center think tank. Start with this fact, which is undisputed: Global heat deaths among older people have increased 68 percent in less than two decades. Now add this qualification: the increase in heat deaths is *caused* by rapidly rising temperatures. With the addition of the qualifier "caused," the contention is transformed from a correlative relationship (heat deaths correlate with age) into a causal linkage: the heat deaths among the elderly are caused by the rapidly rising temperatures.

Not mentioned is the increase within the population of people 65 years of age or older. This has risen 60 percent in the last two decades which, when the increase in mortalities is adjusted for population growth, statistically leads to the conclusion that heat related deaths have risen by 5 percent, not 68 percent! The adjustments of deaths to population-group growth (an increase in this case) is an elementary statistical consideration that for some reason the statistician omitted and a less than prescient editor failed to demand. Widening the scope of inquiry a bit provides some clarity.

At the other end of the extreme from heat deaths are deaths secondary to extreme cold. This wasn't mentioned at all in *The Lancet* article despite a vast disparity: In the United States and Canada between 2000 and 2019, an average of 20,000 elderly people died from heat, compared to 170,000 cold-related deaths.

According to the Lomborg, writing in the *Wall Street Journal*, "Based on today's population size, the current temperatures cause about 17,000 more heat deaths in older people but also result in more than half a million fewer cold deaths."

Obviously reporting on heat deaths but ignoring cold deaths is at the very least misleading. To achieve a valid marker of the effects of rising temperatures worldwide, it's necessary to factor in the increases in population and heat-related versus cold-related deaths.

Nor can any of this be attributed to spin. Taking into account the general increase in population among the elderly, along with the information on the prevalence of cold-related deaths, transforms a frightening statistic (68 percent increase in heat-related deaths among the elderly) to a more manageable one (5 percent).

COLD VS. WARM STATISTICS

Since statistics aims at cold, rational objectivity, sentiment seems to play only a small role. And I'm speaking here of statistical probabilities that are correctly derived and applied. But most of us can bring to mind examples of things that seem statistically obvious to us but aren't accepted as true by others, or vice versa.

For example, several years ago, I was treating a woman in her midthirties for epileptic seizures. The standard medications were not effective in controlling her seizures. She experienced two to three seizures a month. Since seizures can be dangerous and sometimes fatal (a seizure starting when one is at the top of a stairwell or in a bathtub), the patient's seizure control was unsatisfactory.

After I initiated a new anticonvulsant, not only did her seizures cease, but also she experienced an over twenty-five-pound weight loss as a side effect of the new drug. Six months after starting the new drug, my patient was truly a transformed person. No seizures in months, with the side benefit of a slimness that she had formerly cherished but was never quite able to maintain. But at this point statistics entered the picture.

The pharmaceutical company that manufactured the anticonvulsant sent out a warning letter to doctors stating that a few cases of a serious blood abnormality had arisen in patients taking the drug. The statisticians concluded that the odds of anyone coming down with the blood abnormality was one in a hundred thousand people.

I remember clearly sitting with my patient and presenting her with these facts. She listened attentively and asked several pertinent questions and then said, "I don't think I want to continue with this medicine."

I attempted to "reason with her" (my description of our communication): "If I were in your place, I would definitely accept the odds and continue with the medication."

She responded, slightly testily, "I think you've already made it clear what you would do, doctor, but we are talking here what I am going to do, and I don't intend taking a drug that one out of every hundred thousand comes down with this blood reaction." Two months later, she returned to the office. Her seizure frequency had risen to one or two times a month, and she had regained twenty-some pounds.

So, what decision would you have made under these circumstances?

Each of us, even though we may not be conscious of it, has a risk tolerance that may differ markedly from other people, even those who are closest to us. My patient's husband accompanied her to one of the later visits and made the same arguments to her that I had presented. He wasn't any more successful than I in persuading her to restart the medication.

Dilemmas such as this are rarely considered when speaking about the statistics of risk factors. If an effective Alzheimer's drug is eventually developed—and I am confident that it will be—the drug will certainly carry risks, in some cases potentially serious ones. In response, some people with early Alzheimer's may refuse the drug or delay the decision to take it in order to see how things play out over the next year or so. If they delay long enough, they may reach the point when they can no longer give a truly informed consent. What then?

CHRIS HEMSWORTH'S DILEMMA

Imagine your doctor says the following: "We now have a predictive test that serves as an estimate of the likelihood you will come down with Alzheimer's disease by age 65. The test is not 100 percent valid—no test ever is. But if the test turns out negative, there is only a 10 percent chance you will develop the disease. If the test is positive, you will have a nine out of ten likelihood of Alzheimer's by age 65. Would you like to take the test?"

How would you respond to that?

Here is the second related question: if you choose to take the test, would it matter one way or the other that the result will be known only to you? In other words, your doctor will provide you with the test kit and instruct you how to self-administer it. The results will be available only to you. Of course, you can share the results, but you don't have to.

As an alternative, the test can be administered right there in the doctor's office and the result incorporated into your chart. Which would you choose and why?

For the sake of discussion, let's suppose you choose not to take the test because it would be too upsetting if that test turned out positive. Can you foresee that at some later point, pressure may be put on you to take the test and share the results with others? A future marriage partner, for instance, might feel he or she has the right to know the test results to decide whether to have a family or, later in the marriage, to purchase long-term-care insurance.

And if you chose to take the test, how would you respond if it turned out positive? To get more specific, imagine you have inherited a gene, ApoE4, from each parent. The unfortunate result of this genetic assignment makes you more than ten times more likely to develop Alzheimer's disease. In response to that sobering news, would you elect to keep on living your life as you have before, or would you withdraw? A major contribution to your response would obviously involve your financial situation. So, to eliminate that extraneous factor, let's pretend you are independently wealthy, say, a major film actor. Assuming you enjoyed acting, is it likely you would continue your acting career?

Actually, this example is not as arbitrary as it might first seem. In November 2022 the actor Chris Hemsworth, as part of his role in *Limitless*, a documentary series exploring the pursuit of longevity, made a fateful decision. In the show, Hemsworth speaks with experts and explores theories on how humans can live longer, stay healthier, and feel generally good about themselves. As part of the show, Hemsworth decided to volunteer for genetic testing. The episode called for him to learn the results of the testing on camera.

When the Alzheimer's genetic test turned out positive, the doctor and Darren Aronofsky, the creator of the show, canceled the on-camera revelation of the results and spoke privately with Hemsworth. Later, Hemsworth told *Vanity Fair* that the test confirmed his "biggest fear." He announced that he was taking a break from acting and told a BBC reporter that the revelation "triggered something in me to take some time off." Elaborating on this later, Hemsworth indicated he would use the time to take "preventive steps since they can affect the rest of your life."

Let's take a moment to dissect what Hemsworth was really saying, or at least implying. First, he meant the preventive steps require him to devote pretty much all his time to avoiding Alzheimer's disease. None of the preventive steps presented in this book, or any other source that I'm aware of, requires such exclusive dedication to that task. More than likely, Hemsworth's decision was based on his emotional, rather than rational, faculties. This is not a criticism, incidentally. Who knows for sure how any of us would respond on learning what Hemsworth learned?

Current blood testing for the abnormal protein of beta amyloid depends on what's known as an *amyloid probability score*: the probability that sufficient plaque exists within the suspected brain to show up on amyloid imaging study (a PET scan), currently the gold standard test for Alzheimer's disease. The results of the amyloid blood tests agree with the PET scan results about 85 percent of the time. Phrased differently, 15 percent of the time, the results do not match. The soon-to-be-available updated test is expected to raise the percentage to a 90 percent agreement (10 percent disagreement). Indeed, 90 percent is about the degree of accuracy of the PET scan, leaving 10 percent of those tested wallowing in uncertainty—nowhere near the 99.999 percentage discussed earlier.

Welcome to the brave new world of predictive testing. Thanks to advances in genetics and stem-cell research, commonly available and easily administered tests will soon provide predictions about degenerative diseases like Alzheimer's. Indeed, they are likely to become as common as prenatal tests now used almost universally to reveal gender.

THE NUN WHO WORKED IN THE POST OFFICE

Experiencing the onset of dementia isn't like falling down a flight of stairs: unpredictable, sudden, and with maximal damage in close approximation to the inciting cause. It's more like a slow walk in a swimming pool, starting from the shallow end and moving toward the deep end.

This gradual onset of dementia was suggested by an intriguing study of aging and dementia known as the "Nun Study." Started in 1986 by epidemiologist David Snowdon, the study involved 678 nuns in convents across the United States. At the start of the study, some of the nuns were functional and healthy, while others displayed various degrees of disability. Their ages varied from 75 to 103 years old. Snowdon chose nuns for his subjects because, for the most part, they lived very similar lives: free from excessive alcohol use; in the majority of cases, no smoking at all; and living and working in similar environments. Snowdon and his associates were provided access to medical records extending over the nuns' entire lives, along with access to family, medical, and education records. The nun's social and work profiles were also provided to Snowdon.

In addition to these sources of information, Snowdon benefited from something truly unique: copies of the statements each of the sisters had written when they were entering the religious order, usually at about twenty-two years old. These autobiographical essays, written decades earlier, provided Snowdon with the equivalent of an overview of cognitive development starting at entry into the convent and extending to cognitive testing carried out in the 1970s and afterward. While reviewing these autobiographical essays, Snowdon observed that the best functioning nuns differed from their counterparts who had succumbed to dementia by what he termed *cognitive density*: many thoughts and ideas woven into few sentences and paragraphs.

Here is an example statement composed seventy years earlier by a nun, who at the time of her interview at the age of ninety-three had just finished writing a biography and engaged regularly in knitting, crocheting, card playing, and daily walking: "After I finished the eighth grade in 1921, I desired to become an aspirant at Mankato [a convent], but I myself did not have the courage to ask permission of my parents. So Sister Egreda did in my stead."

Contrast this with the statement from another nun, now in her late nineties, showing signs of dementia, who wrote in her early twenties, "After I left school, I worked in the post office."

The first nun presents her vocation as marked by complexity, ambivalence, and perhaps even some unwillingness. She could not bring herself to mention her vocational wishes to her parents.

The second nun, in contrast, leads with only a plain sentence concerning where she worked before entering the convent.

The Nun Study—which has many other fascinating aspects that I'm going to omit here for the sake of making my major point—underscores the long trail of the disease, which may start decades before the affected person shows any objective signs. Thus, the Nun Study adds an additional reason to believe that Alzheimer's disease starts many years before it's first identified by physicians and family.

CHAPTER VIII

THE TWO CRITICAL *PHYSICAL* PREVENTATIVES FOR ALZHEIMER'S

VICIOUS PETS

First a principle: Conditions that increase the chances of dementia also provide methods for preventing it.

Three people are engaged in a lively discussion about pets.

"I'm a dog person. Nothing like a dog for companionship," says the first.

"My mother kept a number of cats around the house when I was growing up. I have inherited her preference," says the second.

"Me, I like *vicious pets*," says the third.

"You prefer vicious pets!?" one of the others says incredulously.

"I didn't say vicious pets. I said *fish as pets*."

* * *

On another occasion the same person experienced a similar hearing error during a discussion with a fellow photographer.

"Let me show you some prison pictures," said the friend.

"Prison pictures? What were you doing taking pictures in a prison?"

"Not prison pictures, but *prism* pictures."

* * *

One final example involves yet again the same person, accompanied by his wife, while touring a museum. At one point they encounter a bronze statue depicting a lion holding a shield.

"Look at the name," suggested his wife. He carefully looked over the entirety of the statue and said, "I don't see any name. It must have been unsigned."

"I didn't ask about the *name* of the artist. I was pointing out to you the beautiful rendering of the lion's mane."

* * *

Such mishearing experiences—despite their amusement value to others—are anything but amusing to the person with the hearing problem. Not only do people with decreased powers of hearing experience a word differently than the speaker intended, but also the whole thread of the communication may be affected.

"Vicious pets" elicits from others disbelief, shock, even a tad of annoyance. The vicious pets, the prison, and the name mishearings lead to erroneous interpretations of the speaker's intention. These episodes are typical of a high-frequency hearing loss of mild-to-moderate degree. Disabling hearing loss can be found in 50 percent of people over 75 years of age. According to a 2020 Lancet Commission on Dementia, hearing loss is the single largest potentially modifiable risk factor for dementia, responsible for up to 8 percent of dementia cases. Please reread that sentence again because it's important.

In a study released in 2023, the National Health and Aging Trends Study (NHATS), which has been ongoing since 2011, found that those with moderate to severe hearing loss over age 65 showed a 61 percent higher prevalence of dementia, compared to those who had normal hearing.

"Age-related hearing loss is the most important modifiable risk factor for dementia in mid-life," according to Madia Lozupone, an international expert on the effects of hearing loss and a faculty member at the University of Bari Aldo Moro in Italy.

In an editorial for *Neurology*, the official journal of the American Academy of Neurology, Dr. Lozupone amplifies further on what she refers to as the *cognitive ear*: "The provocative term 'the cognitive ear' implies that other associative cortical areas process hearing functions in addition to the ear and auditory cortex."

In other words, things *aren't simply misheard but are also misinterpreted*, and these mistaken auditory impressions are incorporated into inappropriate responses: reacting with incredulity to "vicious pets"; attempting to formulate how a photographer friend with no ties to law enforcement wound up taking pictures in a prison; looking for the artist's name on the statue. The results of such mishearing, if continued long enough, can result in cognitive changes. If these cognitive changes are ignored or neglected, dementia may result.

The contribution of hearing loss to dementia varies from country to country. In the United States, the number of cases of dementia solely secondary to hearing impairment is lower than elsewhere in the world. According to the Brazilian Longitudinal Study of Aging, which looked at 9,412 participants, hearing loss was third in overall frequency (6.8 percent), exceeded only by less education (7.7 percent) and hypertension (7.6 percent). Racial and ethnic contributions played a smaller role, and poverty did not greatly impact the results; the richest and poorest regions only differed by about 6 percent. In a preprint of their study for *Alzheimer's & Dementia*, the authors wrote, "Brazil's potential for dementia prevention is higher than in high income countries. Education, hypertension, and hearing loss should be primary targets."

The differences in deafness among countries are partly due to health-care access. Something as simple as impacted wax (cerumen) in the ears is the most common cause of deafness worldwide. The cure consists of nothing more than removing the cerumen from the ear. In India, due to lack of medical access, nonmedically trained workers carry this out (apparently quite successfully) in informal settings.

TRY HARDER

When people are engaged in a conversation under normal circumstances and among people with normal hearing, listening to someone talking stimulates the auditory area within the anterior temporal cortex (ATC). But if the speech becomes permanently degraded by hearing loss, the anterior superior temporal cortex (ASTC) and the posterior superior temporal cortex (PSTC) become activated.

Further, under conditions of speech degradation, areas of the brain normally devoted to attention and other cognitive processes come online. These include the frontal regions, especially the left frontal gyrus. Accompanying the recruitment of these additional areas comes a subjective sense of effort. The person literally has to *try harder* to understand spoken language by concentrating more. This additional effort hijacks the brain areas ordinarily used for evaluating and fashioning a response to what's heard. As a result, the listener with impaired hearing typically functions at a reduced efficiency and misses or may mishear what is said to him, especially the nuances.

To get a feeling for this, imagine yourself learning a new language, say, Spanish. As part of your learning process, you get together weekly with friends whose primary language is Spanish. It's agreed that during the meeting, Spanish alone will be spoken. For you, these meetings provide an extremely helpful opportunity. But they are also demanding and—truth be told—a bit exhausting. You must concentrate really hard on hearing correctly what is said and responding with the meaning you intend. This Spanish-learning exercise serves as a metaphor for what a hearing-impaired person has to go through when listening to someone speaking his or her own language. Mishearing may lead to misunderstandings, which may prove awkward and embarrassing, leading to mild cognitive impairment and, if continued long enough, dementia. Logical thinking may also be affected.

Psychiatric symptoms such as auditory hallucinations may also arise as hearing impairment progresses. Included here are voices or music. The voices most frequently do not convey any content but may. The music isn't usually identifiable but on occasion may be.

If hearing loss eventually leads to dementia, then some degree of comfort can be gained by using hearing aids to counteract this progression. And a rich literature exists suggesting this is an effective strategy. In the interest of being fair, let me make an admission here. The person who experienced the vicious pets, prison, and name episodes was me. Occurring about ten years ago, they provided the first indication that I was experiencing mild-to-moderate hearing loss. They also spurred me to seek help. I now wear hearing aids, and they have made a tremendous difference in my life. I'm placing my bets and hopes that correcting for the decrease in hearing can help prevent early cognitive impairment, which of course can be the prodrome to Alzheimer's.

CHARLES LULLIN AND THE TINY PEOPLE

The second form of acquired sensory deprivation concerns vision. We are primarily visual creatures. If asked whether we would prefer to lose sight or hearing, most people say they would prefer losing their hearing. Actually, loss of sight is less likely to lead to dementia than hearing loss. Many of the captures that keep us linked to other people are auditory ones: detection of emotional tones in other people's voices, sounds of familiar people and places, etc. As Helen Keller phrased it, "Blindness separates people from things; deafness separates people from people." But I don't want to push that point too far. Either the loss of hearing or the loss of vision increases the likelihood that dementia will eventually follow.

In 1760 Charles Bonnet, a biologist and philosopher, noticed that his 90-year-old grandfather, Charles Lullin, was seeing things that were visible to him alone. A few months later, Lullin underwent surgery for cataracts in both eyes. Unfortunately, his visual function continued to deteriorate even further. He began commenting on seeing detailed, miniaturized figures of men, women, birds, various animals, buildings, tapestries, and carriages. Although these hallucinations were concerning to Bonnet, his grandfather was unconcerned about them. He readily conceded that the people and things that appeared to him were not real but products of his imagination. Other than these strange hallucinatory figures, he remained cognitively intact and in good general health without indications of any psychiatric disorders.

Over the next half century, physicians, including neurologists, psychiatrists, and ophthalmologists, made gradual headway toward explaining this strange association between visual loss and visual hallucinations. Their progress wasn't helped at all by the unwillingness of many patients to admit to their hallucinations. But who can blame them? Seeing things that others don't see, things that aren't there, is almost universally considered a sign of insanity. As the condition—dubbed Charles Bonnet syndrome after its original describer—became more readily recognized and more thoroughly investigated, one correlation stood out: the worse the vision, the greater the risk of experiencing visual hallucinations.

Disagreements arose between psychiatrists and neurologists about the origin of the disorder. Despite the almost universal association of Charles Bonnet syndrome and visual loss, many psychiatrists placed their emphasis on the content of the hallucinations. Neurologists and ophthalmologists, on the other hand, downplayed the importance of the content of the illusions and insisted the condition was caused by the affected person's visual compromise or loss.

Today we include Charles Bonnet under the rubric of *acquired sensory deprivation syndromes*. As with the loss of hearing described earlier, damage to vision isolates the brain. Absent normal sights and sounds, spontaneous impulses arise from the visual or auditory centers of the brain in the form of visual or auditory hallucinations.

Those suffering from auditory hallucinations, in contrast to visual hallucinations, often believe in the reality of the things they are hearing. This belief in the reality of auditory hallucinations extends sometimes to paranoia.

But if you think for a moment about the experience of a hearing-deprived person (believing the auditory hallucination) and compare him or her to a vision-deprived person who realizes that the visual hallucinations aren't really there, the discrepancy makes sense. A blind person can easily maintain emotional contact with others because he can hear and speak with them. But deafness robs the affected person of content along with such socially orienting things as tones of voice, rapidity of speech, and emotional resonance—fertile grounds indeed from which suspicion and paranoia may arise.

BLACK PATCH DELIRIUM

In 1958 two physicians, Avery Wiseman and Thomas Paul Hackett Jr., described in the *New England Journal of Medicine* the mental state of patients after eye surgery. At that time, cataract surgery involved the use of eye patches that were applied in the immediate postoperative period. Their paper was intended to call attention to the fact that many of the patients became delirious (a totally reversible form of dementia), losing all sense of where they were, accompanied by visual hallucinations.

As Wiseman and Hackett pointed out, *black patch delirium*, as it came to be called, was not due to the cataracts directly or anything related to the surgery, but rather to the loss of any visual input thanks to the eye patches. In its fully developed form, the patients were disoriented in time and place. They also experienced vivid and frightening visual hallucinations. Since the symptoms were worse at night, the doctors concluded that the normal auditory stimulation available to the patient during the day—secondary to the general hubbub of a busy hospital ward—was absent at night and replaced by the comparative quiet of the nighttime ward. In this setting, absent any auditory or visual stimulation, the temporary loss of vision in both eyes resulted in visual hallucinations.

"Under these circumstances, somewhat analogous to sensory deprivation, misinterpretations may transition to delusions, and anxiety may transform into panic," wrote the doctors. The treatment of this troubling condition was simple: never patch both eyes at the same time. This harrowing experience (for both patient and doctor) is the reason ophthalmologists today never operate on more than one eye at a time. Even when both eyes require the removal of a cataract, the procedure is done by means of two separate operations carried out on two occasions, separated by at least two weeks.

BLIND BUT NOT BLIND

Early in my neuropsychiatric career, I encountered an extreme example of the disjunction that can exist between vision and belief. As a result of a stroke, a 59-year-old woman suffered a total blindness. All the tests confirmed her blindness. If I held a small flashlight before her eyes, she could not reliably identify when the light was on or off. What was unusual and made her stand out from any vision-deprived patient I had ever previously encountered was her assertion that she could see and was not blind. Like many of the other doctors attending her, I spent a considerable amount of time explaining the results of her tests. As our discussion progressed, she became tetchier. Finally, in exasperation she leaped from her bed before I could stop her and hurried across her hospital room where she collided directly with a wall. She struck the wall with such force that it was obvious she never anticipated making contact with it. At that moment, I belatedly realized that the woman was afflicted with Anton syndrome, the delusional denial of blindness.

The first mention of the condition was made by Seneca in his *Ad Lucilium Epistulae Morales*:

> You know Harpaste, my wife's female clown; she has remained in my house. Now this silly woman suddenly became blind. The story sounds incredible, but I assure you that it is true. She does not know that she is blind. She keeps asking the attendant to change her quarters. She says her apartments are too dark.

My patient, as with the female clown referred to by Seneca, exemplifies a body-image disturbance, specifically the absence of knowledge of a body defect, in this case blindness. In this thankfully rare disorder, the statement "I'm not blind" overrides the absence of visual input secondary to blindness.

Charles Bonnet syndrome, black patch delirium, and Anton syndrome highlight the interplay between sensory stimulation (or loss of it) and thinking, especially belief. Loss of vision, when permanent, presents a challenge, because part of the visual pathway within the brain is destroyed (specifically the visual end organ in the occipital cortex at the back of the brain on each side).

Normally, sensation and belief work in tandem, with sensation being the ultimate arbitrator. ("Seeing is believing," we confidently assert.) But on occasion false beliefs can overpower the evidence of the senses. When a person is deprived of either hearing or seeing, beliefs can go wildly astray. Claims that seeing something directly is less reliable than believing something about it are met by understandable incredulity or even mockery, and rightly so.

"Wait a second. I can explain everything," stammers the husband caught by his wife in flagrante delicto. "Are you going to believe what I tell you or your lying eyes?" Well, I think we know which of the two sources of information the wife is going to believe.

CHAPTER IX

PROVEN LIFESTYLE WAYS TO COMBAT ALZHEIMER'S AND OTHER DEMENTIAS

A SURVEY

Several years ago, I did some research into accomplished older Americans who continued to be creative and successful in their seventies and eighties. How did they manage to accomplish this? What were the attitudes and habits associated with successful and healthy brain aging?

Longevity is commonly equated with successful aging. But that isn't necessarily so. Just staying alive without any requirement for quality of life wouldn't be acceptable for most of us. A better measure, it seems to me, are the traits that have contributed to lifetime success even into the eighth or ninth decade of life. To explore this, I interviewed creative men and women in that age group. I asked them what they considered most contributory to their continued success.

Included in my interview cohort were the following:

Morris West (age 80), author of twenty-seven novels and working at the time on the first novel in his *Papacy* series.

Charles Guggenheim (age 71), a documentary filmmaker and winner of four Oscars during his career.

Daniel Shorr (age 80), journalist and prolific commentator for National Public Radio.

C. Vann Woodward (age 88), one of nation's most revered and influential historians, still serving at the time as editor for the *Oxford History of the United States.*

Art Buchwald (age 73), a nationally syndicated columnist and author of dozens of books, including his powerful autobiographies *Leaving Home* and *I'll Always Have Paris.*

Harriet Doerr (age 86), who didn't start writing until age 65 and won the National Book Award for her novel *Stones for Ibarra.*

Sadly, all of these interviewees are now deceased after long and successful lives.

Since that survey, in subsequent years I have added to that list, some on the basis of additional interviews on the topic of healthy brain functioning and how to prevent Alzheimer's disease. Ten personality traits and factors were cited most often by my interviewees as forming the basis for healthy brain functioning, creativity, and a preventative for Alzheimer's disease: (1) education, (2) curiosity, (3) energy, (4) keeping busy, (5) regular exercise and physical activity, (6) acceptance of unavoidable limitations, (7) the need for diversity and novelty, (8) psychological continuity over the lifespan, (9) the maintenance of friends and social networks, (10) the establishment and fostering of links with younger people.

The most frequently mentioned trait by all of those I interviewed was curiosity. Art Buchwald summed up the importance of curiosity in this segment of our discussion:

> To remain creative and mentally sharp, you have to come up with things nobody else has thought of, or you have to deal with familiar things in novel ways. But most important of all you have to have a sense of curiosity. For instance just a few minutes ago while walking here on the Vineyard [Martha's Vineyard, where our discussion took place], I came upon a

minor traffic accident. I stopped because I was curious about the people involved and how the accident happened and why it happened. Other people walk by these kinds of situations and think only about such things as insurance, who was involved, and who might have to pay. But that's no way to remain mentally sharp and mentally vibrant. You have to stay curious about people and what they are doing. Who are the people in the cars? Where were they going? What is their response to this relatively minor mishap? I'm curious about such things. If interest and curiosity stop coming automatically to you then you are in trouble, no matter how young you are. In fact, I believe interest and creativity prolong and enhance life. I believe people who are interested in the people and events around them live better and feel younger.

While thinking about these ten suggestions, you may be able to come up with long-lived people with perfectly normal mental function who serve as exemplars of one or more of the suggestions. My mother, who lived to 95 years old, was intensely curious. I remember that when I was in grade school her reading habit was truly prodigious. She averaged a book a day on a wide range of subjects. But she exemplified most of all the tenth trait on the list: "The establishment and fostering of links with young people."

As she explained to me, "If you want to keep mentally sharp, take the time to socialize with younger people. That way you'll know what's new and fresh."

I asked her, "Don't some young people give you the cold shoulder and want only friends their own age?"

"That's true," she replied. "But you also have to develop a thick skin. Don't take offense. Say something nice about them or their family, and they'll often want to know you better."

Over the years, I've added to that list of traits, some on the basis of additional interviews, some on the basis of lived experience. They include workarounds for age-associated limitations.

Anyone over 55 years old will eventually have to deal with physical ailments that, as a secondary effect, impair thinking. The most common one is impaired hearing. Age-associated hearing loss endangers our capacity to completely relate to those around us.

As hearing difficulty worsens, the affected person may lose the drift of more and more conversations. In response to this, it is only too easy to prefer one's own company or the company of others only under special circumstances (a quiet background, few conversants, and arrangements favoring direct face-to-face discussions in order to facilitate lip-reading). In certain professions, such as actor or therapist—both of whom depend on processing the nuances of every word—this can be the death blow to a career.

The actress Angela Lansbury, who died in October 2022 at the age of 96, required an earpiece during the last few years of her career in order to stay on cue during her appearance in Noël Coward's *Blithe Spirit*, a performance that earned her a fifth Tony Award. She told the *New York Times* theater reporter Patrick Healy, "It's not something you ever want to do. To play important roles at our age, where our names are above the title on the marquee, we are going to ask for some support if we need it." Lansbury asked for such support from the *Blithe Spirit* team, who agreed to her special accommodation. She was also frank and candid with Patrick Healy in describing her workarounds for what she described as a combination of age and memory difficulties.

THE POWER OF NEW WORDS

Earlier in the book, I mentioned memory exercises that will reduce the chances of dementia. Here are several exercises that will strengthen your memory.

Ever since I was about twelve years old, I have always learned at least one new word a day. My father trained me in this practice via a clever inducement. Each day I would fan through the dictionary until I discovered and learned a word that interested me. As an encouragement in this practice, my father often placed a dollar or a five-dollar bill in the dictionary, which was my reward. Sometimes there was nothing there; I never knew until I opened the dictionary. But money or no money, each day I learned a new word.

After learning the word I'd write it down. The result is two shelves of notebooks starting from that period that contain the words that I've learned over the years. When I look up a word from one of the notebooks, I try to extract it from my semantic memory and bring it into conscious awareness. Rarely, I've been able to actually remember the occasion when I wrote a certain word into my notebook. This was an episodic memory emerging from the general background of semantic memory.

Learning new words serves a special purpose later in life. With aging we forget a certain number of words a day. (Nobody knows how many. Because the phenomenon is tough to measure, only estimates are possible.) Think of the new word as a replacement for one or more of the words that are forgotten. Today my new word is *contronym*: "a word that can mean its opposite." For example, cleave can refer to dividing something into parts using an instrument like a cleaver. Or the word can mean adhering, such as when peanut butter cleaves to the roof of the mouth. As a challenging memory exercise, I try to think of other contronyms such as *peruse*, which can mean "to examine carefully" or, its more typical meaning, "to skim." Each of the new words once learned resides somewhere in the vast structure of semantic memory.

Of all the different varieties of memory, working memory is the one you should exercise the most to keep your brain sharp. Let's describe how to do that.

THE MAGIC OF EXAGGERATION

If you exaggerate a mental picture by making it bigger or brighter or louder or sexier, it's easier to remember. Further, endowing these pictures with exaggerated features can exert a powerful effect not just on memory but also but on all aspects of our thinking. These exercises are likely to help prevent dementia.

Try this exercise. Visualize pictures of ten places you frequently encounter along a walk in your neighborhood. Try to see them with the clarity of a crisp photograph. In fact, you can enrich the crispness of your memory pictures by photographing the sights you have chosen and studying them at leisure moments. My "memory pictures" consist of:

1. My house
2. The nearby library
3. A coffee shop
4. A liquor store
5. The front of Georgetown University Medical School (which I attended)
6. The entrance to Georgetown University
7. A well-known restaurant, Café Milano, in Georgetown (my favorite)
8. Key Bridge, connecting Georgetown to Roslyn, Virginia
9. Iwo Jima War Memorial, commemorating the Marines raising the Stars and Stripes on Iwo Jima's Mount Suribachi
10. Reagan Airport

Pick out ten sights that you see every day while walking or driving for your own list.

I use those ten mental pictures as initial placeholders for the items I am trying to memorize. For instance, if I am trying to memorize ten grocery items to buy at the supermarket, I exaggerate each of them some way before placing them on their placeholders. Suppose the list is (1) cereal, (2) hot dogs, (3) coffee, (4) rum, (5) eggs, (6) ketchup, (7) steak, (8) fish, (9) pasta, and (10) watermelon. I place each of these on my memory trail as follows: my house (in the form of a huge box of shredded wheat);

the library (hot dog rolls are lining the shelves instead of books); coffee (people sitting outside my favorite local coffee shop—Black Coffee—drinking from huge coffee cups so heavy, each person has to help another in lifting it); MacArthur Liquor (a giant bar is set up with women in bikinis drinking rum and cokes); Georgetown Medical School (doctors coming in and out while juggling eggs); Georgetown University (the entrance is flooded with ketchup with the students wading knee deep in it); Café Milano (everybody is sitting at tables while holding magnifying glasses and pincers as they try to locate and pin down their tiny steaks—only a joke, since Café Milano is quite generous with its servings); Key Bridge (a giant catfish has hurled itself on to the bridge and cars are running over and tenderizing it between the tires and the road); Iwo Jima Memorial (the Marines are lifting—no irreverence intended here—a box of pasta thirty times the usual length); finally the statue of Ronald Reagan at the eponymous airport, holding out his hand in which is nestled a melon.

If you place the items you want to remember on your own ten place markers, the memorization of the ten items will become automatic.

The memory-peg method is easy to learn and can be practiced anywhere; with regular practice you can speed through the whole process in a matter of a few minutes. Each time it's employed, visual memory, working memory, imagination (those dramatic images), concentration, and focus are enhanced. Each of these processes keeps us mentally sharp.

So please select ten places totally distinct from one another but somewhere in your neighborhood that you encounter daily. Now form a mental picture of each of them. Take a picture with your cell phone of each memory peg so you can study small details and thereby enrich your mental image. Your facility with the method will depend on how frequently you practice it. Any list that you write down, also memorize. When you are in the supermarket, see if you can recall the items before looking at the list. If you can successfully do this, you can feel confident that your memory function isn't significantly impaired and will strengthen further each time you do the exercise.

But I'm convinced that the value of exaggeration isn't limited to memory. Thinking in general will be clearer and more focused by making frequent use of exaggeration. In addition, many of the insights provided by mathematics and statistics can, by an exercise of exaggeration, be gained via imaginative visual recreations without all the complications that often accompany mathematics and statistics. Take Albert Einstein's thought experiments for example.

Throughout his career Einstein—who certainly possessed a sufficient background in mathematics to prove his assertions mathematically—favored what he referred to as visualized thought experiments (German: Gedankenexperiment). A thought experiment is essentially a mental model elaborated within an imaginary scenario. Einstein's thought experiments included accelerating elevators, moving trains, and flashes of lightning.

As an example of how a visual image can convert a pesky mathematical exercise into an aha experience, consider the Monty Hall problem, named after the host of a television program popular in the 1960s and 1970s, *Let's Make a Deal*. In a typical episode, Hall would request a participant to pick one of three doors. Behind one of the doors was a flashy new car, which the contestant could own, if he picks the right door. At this point before the contestant does anything, the host opens one of the doors behind which is no car, but a goat or a mop! This means that the car now has to be behind one of the two remaining doors. The Monty Hall problem rises from the choice the host then offers the contestant: "Do you want to keep your original choice, or do you want to switch to the other door?"

If you elected to stick with your original choice, you are in good company. Mathematicians from George Madison University, the University of Florida, the University of Michigan, and Georgetown University wrote opinion letters explaining the reason the contestants should stick with their original choice. But that choice is incorrect. To prove this requires a bit of mathematical legerdemain, but mathematics and statistics don't have to be relied on at all. An exaggerated picture will do the trick.

Thanks to Professor Michael Starbird, a mathematician at the University of Texas at Austin, I learned a quicker and more persuasive insight into the Monty Hall problem. Imagine instead of three doors, a hundred doors. The same rule applies: the car is only behind one door, leaving ninety-nine doors carless. Pick one. Your odds of being correct are in one in one hundred—not very good. Now the host, instead of opening one door, as was done in the three-door version, opens ninety-eight doors, with none of them showing a car. That leaves two doors: the door you originally picked and one other door. Would you stick with your original choice (one chance out of one hundred that you are correct) or switch from that choice to the last remaining door (one chance out of two that the car is there)? Pretty obvious now, isn't it? By performing the thought experiment of adding ninety-seven doors to the problem, we created an intuitively obvious solution: exaggerate, exaggerate, exaggerate. Exaggerated visual images can vastly increase your mental performances. This improvement comes about via exaggeration. We are primarily visual creatures, and we think clearest when we engage in visual imaginative exaggeration.

REMINISCENCE BUMP

Not all stages of life can be as easily remembered as some others. For instance, most of us can remember very little from our earliest years of life. A substantial improvement occurs for the period between ages ten and thirty, dubbed the *reminiscence bump* by psychologists. Brain maturation undoubtedly plays a big role in the reminiscence bump. During that period, the brain essentially is going through all the changes that distinguish an adult brain from a childhood-adolescent one. Psychological contributors loom large as well.

During the reminiscence bump, we experience the first days of high school, the first experience and perhaps first expressions of our sexuality, the forming in some cases of lifelong friendships, first driving lessons, and so on. A case can be made that these experiences are remembered so well because the majority of them are first-time experiences. Moreover, many of these first-time experiences are later used as comparison points for subsequent similar experiences (the first date serves as the template for all other subsequent dates). Similar templates are based on our first attendance at a rock concert or Major League Baseball game. Because of the reminiscence bump, more of the experiences from that period are recalled and "revisited" during our lifetime than from any other period and are encoded in an autobiographical memory.

TO KNOW YOUR SELF, REMEMBER YOUR SELF

Autobiographical memory (memories related to self and identity) is based on engagement of the hippocampus for the recall of recent memories. This structure disengages when we are attempting to recall distant events that took place five or more years earlier. This explains why people with Alzheimer's can recall the distant past better than things that happened yesterday.

Think of autobiographical memories as a means of enhancing our sense of ourselves. We know this to be true based on some clever experiments involving normal controls and persons suffering from Alzheimer's disease.

In the experiments, the participants were asked to come up with twenty "Who am I?" statements about themselves. After doing this, they were better able to recall autobiographical memories. The recalled names and other details were more specific, contextually embedded (more information describing when, where, and who), as well as more likely to elicit the emotional states accompanying the original experience. While people with Alzheimer's disease performed worse than those with normal memory, they still were much improved from the responses they gave prior to answering the "Who am I?" question.

Impairment in the autobiographical memories for childhood and early adulthood is related to changes in the strength and quality of identity, according to memory researcher Donna Rose Addis, writing in the journal *Memory*. Especially important are memories in the years from ages sixteen to twenty-five—that reminiscence bump again.

According to Mohamad El Haj, a prolific and internationally known memory researcher, "These findings highlight the relationship between self and autobiographical memory in Alzheimer's disease and demonstrate how retrievable information related to the self may influence autobiographical memory."

Several lessons emerge from work on the reminiscence bump and autobiographical memory. Great benefit can result throughout our lives if we exercise our autobiographical memory, especially for the years within the reminiscence bump. During times when nothing in particular is occupying your mind, practice reviewing memories from this critically important memory period. Sharpen your recall by consulting pictures and videos. Most importantly, talk with people you know from this earlier time, and test in a friendly way the congruence of their recollections against yours. Seek information and remembered details that are new to you. Exercises such as these strengthen the sense of self at any stage of cognitive functioning, ranging from superior memory to mild-to-moderate degrees of Alzheimer's.

Reviewed within the context of autobiographical memory, great benefits can be gained when we embark on the occasional trip down memory lane. Only a few years ago, thinking and talking about the past was not considered especially useful and was even considered a waste of time. But comparing the present with what can be remembered about the past can help in recruiting positive aspects of the self. Unfortunately, we can't take for granted this power to mentally incorporate positive aspects of the past within the present and the future. We have to work at it.

PRACTICE MAKES PERFECT: PATHWAYS TO MEMORY IMPROVEMENT

Fortunately, a recent resurgence of enthusiasm for nostalgia is helping people's autobiographical memory all along the continuum from normal cognition to early Alzheimer's disease. *Nostalgia* is defined in Merriam-Webster as "a wistful or excessively sentimental yearning for return to or of some past period or irrecoverable condition."

Such a definition strikes me as unduly critical. Is there anyone who doesn't look back at their past and encounter days that at least in retrospect are more pleasing than anything they are capable of currently experiencing? Nostalgia for these bygone days when things seem generally better than they are today can spur us to make changes and, equally important, experience past emotions that can transfer experiences from the past into the present and, by implication, the future.

Nostalgic exercises have also been shown to reduce physical pain and improve mood by offering the opportunity to see beyond current sadness or regret. Even something as seemingly simple as listening to old music or watching old movies can resurrect some of the feelings that were experienced many years ago. Recall of these positive emotions and experiences enrich autobiographical memory. The goal is essentially to learn from the past by balancing emotions from the past with the forward-thinking priorities of the present and the future.

Let's move on to the ways that memory is altered by Alzheimer's disease.

MEMORY LOSS IN ALZHEIMER'S

Think back to something that happened yesterday. Now imagine the same thing happening next week.

Most people have little difficulty performing such exercises. They can envision a hypothetical future event with the same memory clarity as they can envision the events that happened in the past—not so with people with Alzheimer's.

Due to memory decline, imagining their future is dependent on the retrieval of only a limited repertoire of memories from past experiences, resulting in notable difficulties in remembering the past and imagining the future. As a result, the future is imagined as nothing more than a replay of the past—only bleaker, more featureless, vaguer. This partially explains why people in the throes of Alzheimer's are less engaged. Why rush to experience a future that promises to be only a repetition of the past? Worse, it's a past that lacks color, details, subtlety. Such attitudes influence behavior. The response by a person with Alzheimer's to an invitation ("Let's go to dinner") is often a shrug of the shoulders, coupled with the universal sign of apathy: a lifting of the arms to waist level accompanied by an upturn of the palms. The gesture implies "So what? It won't make any difference anyway." Such apathy is often mistaken for depression, even by experienced neuropsychiatrists. But a person weighed down by apathy isn't necessarily depressed.

Memory loss in Alzheimer's follows a pattern. Typically, the first sign is *anterograde amnesia*—that is, loss of the ability to form new memories. As a result, the same topic may be discussed multiple times in a short time frame, sometimes even within hours, but no recall exists for any of these earlier discussions. On occasion, especially during the earlier stages of the illness, this memory loss for recent events can lead to impatience on the part of a listener, who may assert that the claimed amnesia for earlier discussions is feigned or represents a deliberate act of oppositional behavior: "I can only imagine you *don't really want to do this* because you agreed to it during our earlier discussion only a little while ago."

This loss of anterograde memory also leads to an emphasis during conversations on the good old days that form part of retrograde memory (a superior retrieval for old memories compared to newer ones).

Think of the semantic-episodic balance as a fluid one that over time shifts from dependence on episodic memories (specific unique real-life experiences) to dependence on semantic memory (generalizations referring to a cluster of similar experiences) as a result of a change within the brain.

Retrieval of recent episodic memory always depends on the flow of information to and from the hippocampus—the initial site within the brain where new information is encoded for subsequent retrieval. The hippocampus also just happens to be damaged in the earlier stages of Alzheimer's, preventing or at least greatly interfering with the formation and subsequent recall of new episodic memories. In contrast, older memories are stored in the cerebral cortex in the form of semantic memories. This is much easier to access since it is not ravaged by Alzheimer's until the late stages of the disease. As a result, a person with Alzheimer's can remember the many times he or she stopped for donuts and coffee on the way to work (semantic autobiographical memory), but is unable to recall stopping there earlier today (episodic memory).

As specific memories slip away, *remembering* is replaced by a form of *knowing*. Something cannot be recalled in detail, but the affected person is confident that it happened. The elderly woman affected with early Alzheimer's doesn't remember going to the agricultural fair earlier this summer. But since she had gone for many years, she guesses that she probably went this year as well. If this occurs only rarely, it isn't necessarily abnormal, since studies of normal aging show a lower production of episodic autobiographical memories (the recall of specific experiences, thoughts, occasions, etc.) than semantic autobiographical memories (general knowledge about one's habits and general attitudes). This dependence on general rather than specific recall shifts considerably toward general recall (*generality*) in early Alzheimer's disease.

Such an overdependence on generalization rather than specific memory leads one to always deal with information in need of updating. As a result, the absence of updated information leads to the loss of updated concepts about oneself and current functioning and abilities. The person with early Alzheimer's may exaggerate statements about their past capacities: "I can drive just as well now as I ever did," may be the response when impaired driving is commented on.

At the extreme, all of the impairments accompanying Alzheimer's may be denied, resulting in the condition referred to as *anosognosia* (literally, "without knowledge of" disease; lack of insight or impaired judgment about the illness). What is the basis for this condition? Overgeneralized self-assessment (outdated and incorrect semantic self-knowledge) is given more credence than objective measures of what's happening in the present (weakened episodic memory). When you combine this overreliance on semantic memory with the weakness of episodic memory updating, you have all that is required for an argument. Heated interfamilial arguments are frequently the result of the disjunction between skills that were once possessed but no longer are. The affected person makes claims, in the present, to previously intact abilities that are no longer present.

Let's move on from memory to the other lifestyle measures that may decrease the chances of dementia.

THE FITNESS REVOLUTION

The explosive growth of gyms and health clubs over the past few decades speaks to the popularity of exercise. To Americans between the ages of twenty-five and forty, words like *workout* and *personal trainer* are part of everyday conversation. But it wasn't always so.

In the 1960s and 1970s, for instance, jogging was rare. I can remember from my medical school days the confoundment and puzzlement that many of us felt when observing one of our classmates running along various streets in Georgetown.

"I wonder who is chasing him," someone said with a smile.

"Is he lonely?" I asked, with a fair amount of jest (a reference to the then-popular film *The Loneliness of the Long Distance Runner*).

But nobody held any ill will toward the solitary jogger. We simply found him . . . *inexplicable*. What could justify going out sometimes in the rain, ice, and snow—in order to run through the streets!

Today we recognize the value of a healthy mind in a healthy body. This is what led to the burgeoning of the number of gyms and health clubs. According to multiple surveys and research studies, exercise leads to an increase in overall health and longevity. Not only that, but the brains of exercisers also remain functional for a longer period and perform better than the brains of nonexercisers.

Perhaps you wondered about the claims, increasingly encountered, that exercise helps fend off Alzheimer's and other dementias. What is the basis for such a claim? One explanation holds that the faster heart rate and increase in blood pressure associated with vigorous exercise leads to an increase in the blood and nutrient supply to the brain. While this is true, there are also compelling reasons to believe that our muscles mediate a goodly portion of the beneficial effects of exercise on our thinking.

As a first step, the muscles during exercise produce chemicals that cross what's known as the blood-brain barrier (BBB).

Simply put, the BBB consists of a cellular network of blood vessels and tissue composed of closely spaced cells that control the microenvironment of the brain. The BBB welcomes the influx of beneficial chemicals into the brain and prohibits entrance of harmful ones. Eighty-two percent of the total volume of the gray-matter fiber tracts—passages by which chemicals are distributed throughout the brain—may be modifiable by increases in physical activity. The hippocampus is the main brain target benefiting from exercise. This is important because the hippocampus, as previously noted, is the initial waystation for the establishment of memory.

Regular exercise increases the volumes of both the gray and white matter, leading to stronger functional connections between brain cells and enhanced thinking. Chemicals called myokines, which are released from the muscles during exercise, cross the blood-brain barrier and increase the brain's production of other chemicals that promote growth, plasticity, and the production of new neurons (neurogenesis) in the hippocampus.

Not all exercise is the same. Intuitively, we recognize that running around the local high school track differs from lifting barbells inside the weight room. At the most basic level, exercises differ in duration and intensity, the types and numbers of muscles used, and the source of energy called upon. During aerobic exercises, which are less intense and carried out over a longer time span, the lungs and heart work hard to provide greater amounts of oxygen to the body, including the brain. This oxygen breaks down glucose and then fats to release the energy needed to perform the exercise. Examples of aerobic exercises include walking, running, cycling, swimming, dancing, and jumping rope. Notice that all these exercises involve longer time frames (minutes to hours depending on your physical condition, purpose, time available, etc.).

Anaerobic exercises involve high intensity exercises of short duration, usually seconds to minutes. Because of the short time frame, the energy is drawn from components (principally stored glucose) that are already available in the body. Anaerobic exercises include sprinting, weightlifting, working with resistance bands, and bodyweight exercises (e.g., push-ups, pull-ups, squats).

The best exercise program incorporates aerobic and anaerobic exercises that can easily be combined. After jogging for a quarter of a mile (aerobic), pick up the pace a bit and perform a rapid sprint for a hundred yards or so (anaerobic exercise). This is what happens towards the end of a closely contested marathon. For most of the race the contestants engage in aerobic exercise, but what is called for when approaching the finish line in a tightly contested race is the ability to call on anaerobic metabolism.

Which exercise is the most beneficial to the brain? The only fair answer is *both*, with perhaps a tiny tilt toward aerobic. Regular aerobic exercise reduces the risk of developing dementia such as Alzheimer's disease or a stroke. Regular anaerobic exercise helps to maintain muscle mass. This is especially important in the legs and prevents falls and assists in the maintenance of agility on one's feet. Agility helps to prevent head injury from a potential fall and the onset of mixed dementia, a combination of early Alzheimer's disease with vascular dementia, due to blood clots or small blood vessel diseases.

One other point about exercise applies to those like me who are not especially drawn to it. The distinction between physical and mental exercises is not as clearly demarcated as you might think.

In the 1984 World Chess Championship in Moscow between champion Anatoly Karpov and Garry Kasparov, the match dragged on without a winner for forty-eight games over several months from September 10, 1984, to February 8, 1985. The contest was hyped as the "unlimited match" because, in lieu of a set number of games (the usual rule), the winner would be the first player to win six games. As two of the best chess players in the history of the game, Karpov and Kasparov were so evenly matched that neither could achieve six wins during the several months of play. Most of the games wound up in a draw.

As the matches proceeded, Karpov steadily lost weight and looked ill. As one grandmaster at the match commented, "Karpov looked like death. His skin was just hanging off his bones." In deference to medical advice, the officials ended the match without a clear winner.

Although Karpov's weight loss could have been secondary to a deficient diet (we have no information available suggesting his diet was deficient), another explanation was also possible but seemed unlikely at the time: simply sitting in place for several hours a day in deep concentration had led to a burning of an excessive number of calories. Little was done to decide between these two alternatives until 2018 at the Isle of Man International Chess Championship. The players were monitored for physiological changes. One player, Mikhail Antipov, burned 560 calories during a two-hour match—more calories than are burned during a five-mile run.

It seemed incredible to believe that intense concentration, along with the stress associated with chess matches at the highest level, could lead to more caloric utilization than a whole-body exercise like middistance running.

Recent research suggests the total caloric expenditure is probably closer to 132 calories an hour, the equivalent of 1,188 calories over a nine-hour match.

Over the decades since the 1984 Karpov vs. Kasparov match, chess players competing at the international level have incorporated diet and physical fitness programs to increase the oxygen supply to the brain. This represents a stark contrast to the smoking, drinking, and late-night parties that formerly took place in the international world of high-level competitive play characteristic of the chess world until the end of the twentieth century. Physical fitness and brain performance are now linked like a hand and a properly fitting glove—again, a fit mind in a fit body.

Watching the former chess champion Magnus Carlsen running on a treadmill or playing soccer with his friends, you would be hard-pressed to identify him as the former holder of the World Chess Championship. He now takes for granted something that in 1984 wasn't too widely appreciated: the brain, like any other organ of the body, benefits from—indeed *requires*—physical exercise. The best arrangement is a combination of the two.

"Combining exercise with cognitive training produces significant health benefits compared with exercise or cognitive training alone," according to Eduardo Benarroch, professor of neurology at the Mayo Clinic in Rochester, Minnesota. So while we all wait expectantly and hopefully for the drug that will cure Alzheimer's, exercise is a powerful tool we can employ in the meantime.

"Regular exercise is of fundamental importance because no pharmacologic approach yet available can mimic its multiple beneficial effects," according to Benarroch.

So I guess I owe an overdue apology to my medical school classmate, the lonely jogger, who intuitively appreciated years before most of us the importance of exercise as a means of improving the performance of the whole body, particularly the brain.

DIET: THE NEW RELIGION?

Diet occupies the place in our society held in earlier times by religion. Certain foods are permanently forbidden. In some circles, diet limitations are adhered to with the same intensity an Alcoholics Anonymous member directs toward alcohol (not to be imbibed under any circumstances). In addition to limiting the variety of our diet, this all or nothing approach is one of the reasons diets are so difficult to adhere to. Yes, salmon is a healthy and tasty dinner selection. But every night? What's needed is enough determination to stick to a diet, except for special occasions (that steak or piece of pecan pie) to keep us motivated lest we succumb to nutritional ennui and stop the healthy diet all together.

Prior to any discussion about specific diets, let's consider the effects on thinking of the *nondiet*—that is, fasting. First, I need to make an important distinction. Fasting isn't simply a variety of dieting. Dieting involves a commitment, however tenuous, to eating certain foods and avoiding others to lose weight. A diet, ideally, is practiced daily and continues anywhere from several weeks to several months. Some people even remain on permanent diets. Fasting, in contrast, is temporary, consisting of a certain number of days restricted to the intake of water and electrolytes only. *Intermittent fasting* involves the absence or strict limitation of calories for eight to twenty-four hours, followed by a return to regular eating. Second, fasting isn't necessarily directed at losing weight. We may skip a meal here or there for no particular reason other than the absence of appetite, or we may alter our diet by abstaining from eating as part of a religious holiday.

Since some researchers have suggested the benefits of intermittent fasting for Alzheimer's (more about that in a moment), it's worthwhile to home in on what happens in the brain when we fast.

Following twelve to thirty-six hours of fasting, the body shifts to physiological state of ketosis. This is characterized by a low blood glucose level and depleted glycogen, the storage form of glucose manufactured in the liver. As a result, additional glucose can no longer result from the breakdown of glycogen. A few hours into a fast, the liver's store of glucose is totally depleted. In response, cells elsewhere in the body release fats. The fats travel through the bloodstream to the liver where they are converted into ketones, small molecules of fat that our body cells use as an alternative energy source when glucose isn't available. This *metabolic switch* as it's called—replacing glucose with ketones—happens after about twelve hours without of food.

While the liver is the primary site of ketogenesis, brain cells called astrocytes can also generate ketones. Within hours of initiating a fast, ketones become the brain's preferred fuel source, providing up to 70 percent of its energy requirements. Ketones are a more efficient source of energy in muscles and probably the brain as well, leading to increased brain cell bioenergetics and cognitive performance. A ketone diet is high in fat, low in carbohydrates, and minimally adequate in protein.

In animal research, rodents on ketone diets for five days improved their performance in spatial learning and memory. Alzheimer's disease typically involves the same improvements in the neurons' bioenergetics, glucose metabolism, and neuronal signaling.

The diet mimics the effects of fasting by generating ketones and bringing about most of the effects induced by fasting. What are the effects on the brain of the metabolic switch?

Many people describe a kind of "ketone high": mood elevation and a feeling of psychological expansion. Some people even feel more grounded and spiritually awakened—perhaps one of the reasons the practice of fasting has been integral to religions throughout history. Most importantly, fasting enhances thinking, learning, memory, and alertness. The basis for this is the stimulation brought on by fasting of a protein in the brain cells called *brain derived neurotropic factor* or BDNF. This chemical exerts its greatest influence on the hippocampus, where it generates new nerve cells—the basis for its stimulating effect on memory. Fasting also triggers a process called *autophagy*: destruction or removal of damaged and dysfunctional neurons.

Intuitively, it would seem to make sense that a ketogenic diet should be helpful with Alzheimer's, where brain glucose uptake is moderately impaired but ketone utilization is not impaired. So far, there have not been sufficient studies to firmly establish the benefit of the ketogenic diet. But in at least one study, fifteen people with mild to moderate Alzheimer's disease reported some improvement in their cognition after twelve weeks of a ketogenic diet.

In all fairness, it has to be pointed out that the slowing of brain cell degeneration induced by fasting has only been proven effective in animals. Nor should this be surprising. Fasting involves a grueling, demanding regimen that not everyone is willing to undergo. Have you ever been deprived, for one reason or another, of food for two days? If you have, you undoubtedly can remember the hunger pangs. But some people would be willing to undergo that temporary discomfort in the interest of decreasing the odds of coming down with Alzheimer's.

In early 2023 a team from Columbia University's Robert N. Butler Aging Center and Mailman School of Public Health showed that fasting in humans slows the pace of aging. Fasting led to a 10 to 15 percent reduction of mortality risk, an effect similar to a regular smoker who quits smoking. Since the human lifespan is so long compared to mice and other animals, direct measures of human lifespan are impractical. So instead, the researchers concentrated on biomarkers developed to gauge the rate and progress of aging during the duration of this study.

As one of the researchers put it, "The fasting diet has been shown to be associated with reduced risk for heart disease, stroke, disability, and dementia."

Because certain medical conditions preclude fasting, it would be advisable to check with your doctor whether fasting is for you. On the whole, fasting appears to be one of several approaches mentioned in this book for decreasing the likelihood of dementia or preventing it altogether. You can expect to see additional studies within the next year or two of the positive effects of the ketogenic diet.

LITTLE WHITE LIES

Diets are fickle. The latest fad-of-the-month diet is succeeded by another that contradicts the tenets of the first one. Take the following example:

1. Eggs are good for you because they provide in a small package a plethora of necessary proteins and vitamins.

2. No. Eggs are bad for you because they are rich
   in cholesterol.
3. No, wait. The amount of cholesterol in our diet doesn't
   necessarily affect the levels in our blood. So eggs are really
   good for you.
4. Not if eggs are linked with heart attacks, which research
   from 2019 points to.

A bit confusing, right? So, should you eat eggs or not?

Unfortunately, nutritional science has several built-in lim-itations that work against reaching firm conclusions about what might be the best diet for lessening the chances of Alzheimer's.

For one thing, we can force lab animals to eat foods we choose for them, but we have to persuade people to keep them-selves on the diet we are interested in evaluating. Second, the lifespan of lab animals is limited to only a few years, while our lifespan is seventy years or more. Moreover, conclusions about the effectiveness of a diet over a lifetime requires proof extend-ing over two or three generations. This isn't a problem in lab ani-mals, where three generations live out their collective lifespans in less than ten years. But to reach the same degree of certainty in our own species would require more than two hundred years of follow-up.

These factors relegate to mere opinion any claim that a diet will prolong life or definitively eliminate a certain disease. Rather, conclusions are reached on the basis of biomarkers: substances in the blood or brain that are currently considered indicators of present or future disease. For instance, a diet that reduces choles-terol will likely be touted as extending life. Since high cholesterol is associated with heart attacks, low cholesterol, it is assumed, will reduce heart attacks and therefore prolong life. Despite this assurance, not one random trial has shown that giving up red meat or avoiding saturated fat will guarantee prolonging *your* life.

The period of observation required to prove such an assertion is simply too long. The same principle holds when it comes to demonstrating the harmful effects of a poor diet. Diets, even seriously faulty ones, take years to affect health.

In addition, the recordkeeping in humans falls far short of the meticulous recording of dietary intake in laboratory animals. Instead, we keep food diaries detailing what we are eating while we are on a certain diet. But if we don't seem to be losing any weight and are frustrated and embarrassed about it, we can scribble little white lies into the diary that will lead the nutritional scientist astray.

Further, dietary studies cannot easily be assimilated into the most effective observational procedure: the clinical trial. Typically, in a clinical trial, half of the participants are placed on the active drug, while the other half is given a placebo. If the cohort on the active drug does significantly (statistically) better health-wise than the placebo group, that finding can be used to buttress the argument that the active drug is effective. But such an approach isn't and can't be applied to human dietetics. Few volunteers will be amenable to sticking to a diet from the time they are old enough to consent to a clinical trial until the end of their life.

Finally—and I believe most importantly—willingness to go on and stick with a diet is usually linked with other behaviors carrying health consequences. Fast food, ultraprocessed food, and food rich in saturated fats and other components of "junk food" are usually associated with poverty and unhealthy lifestyles such as smoking, excessive drinking, drug use, and poor exercise habits. At the opposite extreme, a commitment to a healthy diet is favored by the wealthier and better educated who also engage regularly in challenging themselves physically (increased exercise) and mentally (building up mental agility). I suspect these factors are more important than diet in preventing Alzheimer's or other dementias.

THE SEARCH FOR THE PERFECT DIET

When it comes to diet, avoiding certain foods has proven more effective on improving brain function than selecting and sticking to a particular diet. Let's start with the greatest offender, processed food.

Entirely avoiding processed foods would be unnecessarily demanding. What's more, greater harm than good may result. By definition, a processed food is one that has been altered from its original form. This doesn't necessarily involve anything harmful. Pasteurizing, heating, canning, refrigerating, and drying are all forms of processing. Of course, processed food can be avoided if you pick and rapidly consume your own fruits, slaughter your own beef, milk your own cow, etc. And in addition, processing confers benefits: frozen or refrigerated vegetables are no less nutritious than vegetables eaten soon after picking, and they last much longer. In short, simply because food has undergone processing doesn't imply the food is any less nutritious or healthful.

Ultraprocessed foods are another matter. Technically defined as industrial formulations with five or more ingredients, ultraprocessed foods are converted from processed foods during the final stages of food production, so called tertiary processing. At this time sugar, salt, oil, and fats are added in the general interest of taste and preservation. What's the quickest, surest way of distinguishing a processed from an ultraprocessed food? A long list of ingredients is an indicator that you are dealing with an ultraprocessed food.

Picture someone munching on a chocolate cookie that tastes almost as good as the ones his mother made for him as a child. That cookie, however, includes ingredients unlikely to have been present in mom's chocolate cookie: heavy doses of invert sugar, salt, and gelatinized wheat starch, along with natural and unnatural flavoring. The cookie weighs under two ounces but contains 270 calories (more than 10 percent of a nutritionally balanced diet of 2,000 calories), 7 grams of saturated fats, 190 milligrams of sodium, and 20 grams of sugars. Before we even get to the effects of these substances on the brain, it's well known that excess sugar, salt, and oil play prominent roles in the development of diabetes, heart disease, cancer, and obesity. All of these are contributors to vascular dementia.

Among the ultraprocessed foods (high in added sugars, fats, and salts and low in protein) are soft drinks, salty and sugary snacks, ice cream, sausage, bacon, deep-fried chicken, ketchup, mayonnaise, prepackaged soups, sauces, frozen pizza, and ready-to-eat meals. Added to these are the so-called pleasure foods: hot dogs, sausages, burgers, French fries, donuts—to name only some of the common ultraprocessed foods.

So why haven't most of us committed ourselves to eliminating these items from our diet? That's an easy question to answer. These foods are in sync with our frenetic fast-paced lives requiring us to eat quickly within shorter and shorter time frames. Ultraprocessed foods facilitate such dietary styles. So it is not surprising that 58 percent of the calories we Americans consume come from ultraprocessed foods. Avoiding them entirely is inconvenient, time consuming, and expensive. But if we want to decrease our chances of dementia, totally eliminating them is the place to start.

"WHAT'S GOOD FOR THE HEART IS GOOD FOR THE BRAIN"

In the 1950s, researchers around the world undertook an ambitious and arduous search for the diet that stood the best chance of reducing heart disease. For several decades they minutely examined the diets of thousands of middle-aged men living in the United States, Japan, and parts of Europe. Their goal in this Seven Country Study, as it came to be known, was to identify the diet most effective in reducing the odds of cardiovascular disease.

Early in their studies, the researchers noted an association of cardiovascular disease with saturated fats and cholesterol.

One observation stood out: those who lived around the Mediterranean Sea experienced significantly fewer cases of cardiovascular disease. Their diet consisted for the most part of fruits, vegetables, whole grains, nuts, seeds, legumes (soybeans, chickpeas, peanuts, etc.), and lean protein, especially fish and unsaturated fats. Among those who stuck with such a diet, blood pressure and cholesterol tended to be lower, and type II diabetes rarer. Foods high in saturated fats, such as red meat and butter, were infrequent in this diet and were typically replaced by fish, such as salmon, tuna, and sardines, which are rich in omega-3 fatty acids. Eggs and dairy products (milk, cheese, etc.) were allowed, in moderation. Alcohol in small quantities (typically a glass of wine per day) was included in the aptly titled Mediterranean diet.

A twenty-five-year study published in 2018 found that people (in this case mostly women) who stuck with a Mediterranean diet for up to twelve years experienced a 25 percent reduced risk of developing cardiovascular disease. This was thought to result from better blood-sugar control and reduced inflammation and body mass index (essentially reduced obesity). At the molecular level, a Mediterranean diet led to a reduction of oxidative stress, which produces the DNA damage known to be associated with neurological diseases and cancer.

In addition to the foods mentioned above, I'd recommend foods that are regularly suggested for people with lowered levels of immunity. If immune levels trend low enough, an autoimmune disease can result (rheumatoid arthritis, multiple sclerosis, type I diabetes, Guillain-Barre syndrome, and others). But we are speaking here of low immune levels that don't always lead to an autoimmune disease but probably increase the chances of coming down with one. As I'll discuss in more detail, inflammation plays a role—perhaps a major role—in dementia. This is the reasoning behind incorporating antioxidant foods into your diet. The inflammation seen in Alzheimer's and perhaps the other dementias, incidentally, is internal in origin and not accompanied by the common visible signs of inflammation: redness, swelling, etc.

Antioxidants like vitamins C and E, selenium, and pigments within fruits and vegetables (carotenoids like beta-carotene, lycopene, and lutein) oppose the inflammatory process.

Vitamin C can be found in most fruits and vegetables, with extra portions in citrus fruits (oranges, grapefruit), strawberries, broccoli, spinach, and kale.

Vitamin E sources include nuts, especially almonds, peanuts, and avocados and salmon.

Look for carotenoids in carrots, squash, and sweet potatoes.

Selenium is found in tuna, shrimp, chicken, eggs (see they can be good for you!), oats, and lentils.

In addition to these, consider adding vitamin D. In one study from 2022 published in the *British Medical Journal*, vitamin D supplementation can lower the risk of an autoimmune disorder by 22 percent, suggesting a benefit as well in healthy individuals.

Vitamin D is found in fatty fish (salmon, herring, tuna), egg yolks (those eggs again!), fortified milk and orange juice, and fortified breakfast cereals. One must not forget either that sun exposure leads to vitamin D production through the skin. Unfortunately, the increase in small particles in the air when combined with smog lessens the contribution of sunlight. This is best handled by a judicious use of supplements.

If the vitamin B12 level falls low enough, the end result can be dementia, psychosis, or a combination of the two. Along with alcohol abuse, pernicious anemia is the next most common cause, affecting 0.1 percent of the general population and ten times that number in people who are sixty years of age. Foods with the highest ratio of vitamin B12 include salmon, beef, poultry, and—you guessed it!—eggs.

Iron is important to deliver oxygen to the brain. Rich food sources include beef, pork, shrimp, spinach, and dried fruits. Vegetarians and vegans are faced with a special challenge when it comes both to iron and vitamin B12. Many of the foods rich in these nutrients (vitamin B12 and iron) are meat, poultry, and fish. In extreme cases, vegans refuse to eat any of these foods. Vegans totally refrain from meat. In addition, they do not eat dairy, eggs, or any products of animal origin. Veganism is a purer, more radical form of vegetarianism that, according to one's point of view, comes with a wheelbarrow full of social and political agendas having nothing to do with nutrition. In any case, both vegetarians and vegans will in most cases require supplements of iron and vitamin B12.

Finally, here is a list of the most nutrient-rich foods. They should form an important part of any diet aimed at lowering the chances of brain diseases, such as Alzheimer's and the other dementias: salmon, blueberries, kale, garlic, shellfish (barring any allergies), and dark chocolate with high cocoa content. ("Yum, yum," writes this mild chocoholic who regularly eats one piece of dark chocolate daily.) Two other foods I hesitate to hype too much, because many people find them unappealing, are liver and seaweed.

I think we answered the question raised at the beginning of this section about the nutritional value of eggs. Most dieticians recommend them at least in moderate quantities.

Here is a key point to keep in mind when evaluating the evidence pro and con on the protective effect against Alzheimer's exerted by diet (or exercise or other behavioral interventions, for that matter). Longevity is not a guarantor of retained mental functioning. A centenarian may well be incapacitated by Alzheimer's or other dementias. Indeed, if one in ten Americans over the age of 65, and four in ten in the decade between

80 and 90 years old, are afflicted with Alzheimer's, the likelihood is actually quite high that a centenarian may have Alzheimer's. This has obvious implications, especially when it comes to diet. The Mediterranean diet is associated with good heart health and prolonged longevity. But does it protect against Alzheimer's?

Research released in late 2022 suggests that a healthy diet, such as the Mediterranean diet, *does not* reduce Alzheimer risk. Yes, you read that correctly. In this study, 28,000 adults (slightly more women than men) who were free of dementia were followed for more than twenty years as part of the Swedish Malmo Diet and Cancer Study. Of these, 1,943 people (6.9 percent) developed dementia. Although these subjects were older and less educated and suffered from cardiac disease or risk factors leading to it, diet played no part.

Two unexpected surprises were found: First, individuals who kept to conventional healthy dietary recommendations did not show lower levels of developing dementia. Second, even a more specific health diet such as the Mediterranean diet did not appear to lower the risk of dementia.

To make matters even more disappointing, no association was found between diet and the presence of Alzheimer's-related changes—the amyloid plaques or neurofibrillary tangles. Nor did examination of cerebrospinal fluid (CSF), the fluid that bathes the brain, show any difference between those on a special diet and those who were not.

Should these surprising, even bewildering findings overturn our traditional beliefs about diet? I don't think so. This was only one study. The results may yet turn out to be a false outlier—a value that lies outside the findings of the majority of other researchers studying the nexus between diet and dementia. In other words, the findings could be an error. But whatever the cause of these perplexing findings, it suggests that although you may be able to eat your way toward Alzheimer's by consuming an extremely ill-chosen diet, you won't be able to rely on any diet *alone* to prevent Alzheimer's in isolation from other factors.

As Nils Peters, MD, of the University of Basel, Switzerland, and Benedetta Nacmias, PhD, of the University of Florence, Italy, commented on the Mediterranean diet outlier findings in an editorial in the journal *Neurology*,

> Diet as a singular factor may not have a strong enough effect on cognition but is more likely to be considered as one factor embedded with various others, some of which may influence the course of cognitive function. Diet should not be forgotten and it still matters, but should be regarded as one intervention with respect to cognitive performance.

A PAUSE FOR CAREFUL THOUGHT:
COFFEE, TEA, OR BOTH

Both beverages are looked upon with suspicion. Tell a cardiologist that you are drinking more than two cups of coffee or tea a day, and you'll likely unleash a mini lecture on heart attacks and cardiac arrhythmias. And there is certainly good reason for this concern. According to a paper published in the *Journal of the American Heart Association* on December 21, 2022, drinking two or more cups of coffee daily was found to double the risk of heart death in people with severe hypertension. The operative word here is *severe*. People with mild high blood pressure not considered severe did not show the effect of increased death from cardiovascular disease. The statistics in regard to the brain are more reassuring.

According to recent research, both coffee and tea are associated with a lower incidence of dementia. What some would call "heavy" or "excessive" consumption (two or more cups of coffee a day) resulted in a lower incidence of dementia. Within the Japanese Murakami cohort of 13,826 participants, the men and women 60 years of age or older benefited the most from drinking three or more cups of coffee per day. This finding was entirely consistent with a November 2021 study by the UK Biobank, which analyzed coffee and tea consumption as related to dementia risk. Among the 365,682 participants in that study, those who drank two to three cups of coffee per day and two to three cups of tea per day lowered their dementia risk by 28 percent.

An Australian study published in *Frontiers in Aging Neuroscience* found that overall, those who drank two to three cups per day of coffee showed less cognitive decline and less cerebral amyloid-beta accumulation over 126 months of observation. (Tea consumption was not measured.)

Why may coffee and tea prove helpful in preventing or delaying dementia? For one thing, coffee's active component, caffeine, influences several functions that oppose Alzheimer's.

In the Alzheimer animal disease models, caffeine enhances working memory, alertness, spatial learning, and object recognition. It also reduces amyloid plaques and increases their clearance from the brain. While coffee is clearly associated with a decrease in dementia (consuming more than three cups per day was associated with 50 percent reduction in dementia risk), this is a bit disappointing to me since, although I drink both, I have a slight preference for tea at any time other than breakfast. But as noted by the UK Biobank study, the benefits of coffee and tea can be additive.

Perhaps the difference can be attributed to the extraction process. Green tea is typically extracted several times from the same tea leaves with the higher number of extractions varying with the desired strength of the tea. Thus, more variations exist in the concentration of caffeine in tea when compared to coffee.

Since these findings on coffee and tea are quite dramatic and at odds with traditional thinking about caffeine, I suggest you speak with your doctor before increasing your tea or coffee consumption, especially if you have poorly controlled or complicated hypertension. Such a discussion is even more advisable if you are a noncaffeine person and are considering starting coffee, tea, or both for the first time in your life. Making such a change for a long-term benefit—decreasing your chances of dementia—probably isn't advisable if you have cardiac or other conditions that may be adversely affected by coffee or tea. This is an individual decision that must be based on each person's overall health profile, risk tolerance, and medical advice. For one thing, people vary in terms of how much caffeine they can consume before their normal sleep cycle is disturbed.

THE SELF-FULFILLING PROPHECY

So far no one has come up with a formula for getting a perfect night's sleep. Part of the difficulty stems from failures by sleep researchers to reach agreement about what that would entail, or even how to experience it. As a result, sleep involves a self-fulfilling prophecy. The more anxious about sleep we are when we are going to bed, the less successful we will be in falling asleep. A 2013 study in *Nature and Science of Sleep* found that this anticipatory anxiety (How can I be sure I'll fall asleep tonight? Stay asleep? Feel rested in the morning?) is at its worst during periods of stress. Fueling the anxiety about sleep is the ever-present "advice" offered by "sleep experts," who inform us of such misguided notions as an absolute need for eight hours of sleep, the need to wake up at the same time every morning, and the need to avoid getting out of bed more than once for a bathroom visit during the night.

Sleep problems become more prominent with aging. Both the quality and structure of sleep are affected. Older people suffer from a disturbed sleep triad: it takes them longer to fall asleep, they wake up more frequently during the night, and they feel unrested during much of the next day. They also spend less time in deep sleep. This is the sleep phase when bone and muscle growth and repair occur, when the immune system is strengthened, and—most important for our purposes—when memories are consolidated.

From the strictly behavioral point of view, most of those in their mid-sixties sleep about two hours less than they did when they were in their forties. We have here a paradox: the older person spends more time in bed trying to get to sleep, and the result is less sleep and more dysfunctional sleep (decreased slow-wave sleep).

If you are bothered by any of these sleep problems, a daytime nap can prove helpful in regulating your nighttime sleep. In a study published in the *Journal of the American Geriatrics Society* (February 2021), researchers at Cornell Medical College concluded that napping increases an individual's total sleep time at night. This is important since naps of proper duration and timing should not lead to daytime drowsiness or interfere with nighttime sleep.

The benefits of a brief (five to fifteen minutes) nap are immediately noticeable and last one to three hours. Longer naps (greater than 30 minutes) sometimes result in sleep inertia—waking up from a nap feeling groggy. This lasts only a few minutes, followed by improved cognitive performance lasting for a longer period (several hours). The sleep inertial effect can be quickly dispelled by taking about 100 milligrams of caffeine (about one five-ounce cup of brewed coffee), stepping out into bright sunlight, or washing the face with moderately cold water. Another strategy is to take the caffeine *before* the nap. Caffeine takes about thirty minutes to exert its full effect, which is the perfect duration for a nap. The resulting alerting benefits of the nap can thus be augmented by the alerting effects of caffeine.

I've cultivated the nap habit over four decades and can fall asleep promptly and awaken feeling refreshed twenty to thirty minutes later. In order to adopt the same habit, keep in mind that you cannot *force* yourself to go to sleep. The harder you try, the less successful you will be. This is the well-known paradox of sleep: the greater the conscious effort to fall asleep, the less satisfactory the result.

During the first few days or weeks of trying to nap, simply lie down on a couch at a set time for thirty minutes with no other purpose than to simply mentally decompress. Don't allow yourself to daydream about home or career issues, news events, or anything else that can interfere with your relaxation. After only a few days of this, you will find it easier to gauge when the thirty-minute timeslot has ended. This provides the template for your brain to awaken you, once you have fallen asleep, after about thirty minutes. With continued and steady dedication, you will be able to nap at almost any time you choose. The best time to nap? Within an hour or so of lunch when, as a result of the intake of food, primarily carbohydrates, we are most inclined to drowse off.

In addition to energy restoration, other cognitive benefits will accompany the nap habit.

"Memory improvement is the strongest evidence of a nap's benefits in cognition," according to sleep and nap researcher Rebecca Spencer at the University of Massachusetts Amherst. "Naps benefit various forms of learning, from learning of simple words, to motor learning, to emotional learning."

Numerous laboratory studies have confirmed that naps solidify already learned information. When we first learn something, that knowledge goes into the hippocampus, the brain region responsible for the initial formation of a memory. When we nap, hippocampal activity matches the pattern of activity that occurred when we learned the new information. This is called *neural replay.*

"The brain essentially replays your memories, or a 'movie' of your day," according to Spencer.

SO LONELY YOU COULD DIE

"Does anyone live with you?"

"No."

"Do you have any pets?"

"No."

"That must be very lonely?"

"I didn't say I was lonely, doctor. I said I lived alone."

This somewhat terse exchange occurred at the start of a patient evaluation. Upon hearing of the patient's living arrangements, I rather untherapeutically voiced my own feelings about living alone. I would be lonely. She was not.

Loneliness is an unpleasant feeling that isn't the same as social isolation. As my patient intimated, social isolation doesn't automatically segue into a constricting state of loneliness. Often described as a "dull ache," loneliness results from the risk or merely the perceived risk of social isolation. Genetic factors undoubtedly play a role here. Even in early childhood, certain people prefer their own company to sharing their time with others. Not everyone experiences loneliness under conditions of social isolation. In fact, social isolation isn't even necessary to experience loneliness.

More than one-third of adults ages 45 or older feel lonely despite frequent interactions with other people. While loneliness only might increase with age, social isolation does. Nearly a quarter of adults ages 65 and older live alone due to the death of a spouse or chronic illness, among other factors.

Those who experience this deep ache would be correct in interpreting it as a clarion of doom. High rates of depression, anxiety, and suicide regularly accompany loneliness. I suspect loneliness also plays a role in murders and even in mass shootings. We read frequently in the papers or hear news reports of murders following the breakup of a romantic relationship. Conceivably, the rejected partner may well have reacted to this breakup with a return to an aching loneliness experienced prior to forming the romantic relationship. Such feelings would be expected to set off even deeper pain when the partner has broken up the relationship in order to enter into another one. Several mass murderers have written accounts on social media of feeling rejected or being unable to form close relationships with others. Granted, feelings of rage and perceived rejection, coupled with the wish to destroy people who were the outlet for that rejection, may play a stronger role than loneliness.

Recently, the high cost of social isolation when not necessarily accompanied by loneliness has captured the attention of social scientists and neuroscientists. Their findings are grim. Social isolation alone significantly increases the risk of premature deaths from all causes, a risk thought to be coequal in importance with smoking, obesity, and physical inactivity. But for our purposes, the most sobering risk of social isolation and loneliness—considered together or separately—concerns the risk of dementia. Loneliness increases the risk of dementia by 50 percent, according to a paper published by the International Academy of Sciences.

HIKIKOMORI

Loneliness and isolation are especially prominent in Japan. *Hikikomori*, the Japanese word for "pulling inward" or "being confined," describes modern day hermits and recluses, average age 37, who maintain almost no social contact at all. The chronicity of the self-isolation is especially striking; "prolonged and severe social withdrawal for a time period of at least six months" is one definition of hikikomori. Thanks to smartphones, food delivery services, and internet dependency (if not outright addiction), it's possible in Japan and other technologically advanced countries to live your entire life within the confines of your own apartment.

Among the older population, social isolation and loneliness are common, with 15 percent going for days without any social contact and 30 percent with no one that they can count on.

While some speculate that such a self-imposed isolation represents a serious mental illness, the explanatory flow most probably works in the opposite direction: prolonged isolation leads to increasing loneliness, depression, and even suicidal impulses. How prevalent is hikikomori? One community-based survey places the prevalence at slightly over 1 percent of the entire Japanese population.

Some of the contributors to hikikomori undoubtedly involve the widespread breakdown of social cohesion, disruption of traditional family patterns, and dizzyingly rapid technological progress. Although hikikomori is most prevalent in Japan, social isolation is increasing in many parts of the world. In South Korea, Hong Kong, the United States, and Mexico, increasing segments of the population are choosing to go it alone.

Continued

In a 2020 report from the National Health and Aging Trends Study, investigators found that 24 percent of adults age 65 or older (approximately 7.7 million people) were socially isolated; 43 percent of those over 65 years of age reported feeling lonely. Dropping the age requirement to anyone over 45 years of age revealed 35 percent of the group reported feeling lonely. Most worrisome in all of this is one simple cruel fact: in this study, loneliness increased the likelihood of developing dementia by at least 30 percent.

A study from the UK Biobank, enlisting five hundred thousand individuals from twenty-two centers around the United Kingdom, aimed at distinguishing social isolation from loneliness.

Social isolation was assessed by three questions: (1) "Do you live alone?" (2) "What are your social contacts?" (3) "How often do you get together with other people?" Loneliness was gauged by two questions: (1) "Do you feel lonely?" (2) "How frequently do you confide in close people?" Results? Dementia was significantly higher in people who were socially isolated. Loneliness too led to an increase in the incidence of dementia, but the biggest contributor to loneliness was depression. In other words, social isolation alone was sufficient to bring about structural brain changes, such as lower volumes of the gray matter in the temporal, frontal, and hippocampal regions.

Another research project, the Harvard Study of Adult Development (HSAD), started eighty-five years ago, has come up with some more specific findings about isolation and loneliness. According to Robert Waldinger (director) and Marc Schulz (associate director) of the HSAD, "Over and over again one of the participants in the Harvard Study who reached their 70s and 80s would make a point of saying that what they valued most were their relationships with friends and family."

Although the Waldinger-Schulz findings concentrated on happiness, rather than dementia per se (their book is *The Good Life: Lessons from The World's Longest Study of Happiness*), their findings are directly relevant. Chronic loneliness increases a person's thoughts of death in any given year by 26 percent. Further, loneliness is associated with greater sensitivity to pain, suppression of the immune system, compromised brain function, and disturbed sleep. All of these are contributors to dementia.

So think of loneliness as the brain crying out for social connection. Indeed, the lonely person shares many of the same factors that affect the isolated neuron. Both people and neurons are designed to function at their best in a setting of communal interaction. Both the isolated neuron and the isolated person are destined to die off in the absence of communication with others of their kind.

According to Lawrence Whalley, MD, writing in *Neurology*, "Older people living socially isolated lives are more likely to be lonely, but loneliness alone is insufficient to increase their dementia risk. . . . It is social isolation acting alone and not loneliness that increases the incidence of dementia."

Dr. Whalley's findings make short work of the claims, increasingly voiced by young adults within our population, that social isolation can be freely chosen without any dire consequences— no need to accommodate to anybody's wishes other than your own. Sounds good for the confirmed misanthrope, except for one thing. It's not true. Social isolation is sufficient to bring on dementia—no one to talk to, no one to turn to, no one to love and be loved in return. Sounds like one of those country-and-western ballads, doesn't it? But loneliness is not just the mainstay of country music. It's real-life pain for many people, especially for those sixty years of age or older.

So, no matter the strength of your lone-wolf tendencies, force yourself, if necessary, to regularly engage with other people. Gathering together in hobby clubs or eating clubs every week or so, for instance, may be sufficient. Even limited social engagements hold promise of meeting new people, special projects, and unexpected experiences: new friendships, perhaps even a new romance.

LE MIEUX EST L'ENNEMI DU BIEN

It's best to eliminate all factors from our lives that are known or suspected to contribute to Alzheimer's. Hard to disagree with, right? I guess we can all drink to that! But wait. Drink what?

Alcohol occupies a privileged position when it comes to dementia. Of course, too much is too much: more than two drinks a day for men, one drink a day for women, according to the US Dietary Guidelines. An alcohol abuser conforming to the technical term "excessive alcohol abuse disorder" probably *spills* that much alcohol in a day. Many people who exceed the two drinks (men) or one drink (women) per day consumption don't feel they have a problem unless they experience blackouts, regularly slur their speech while drinking, or repeatedly hear general but firm requests (more like commands) from family and friends to cut back on the booze.

Complicating any attempt to describe who can drink and how much are several common myths about alcohol's effects. "While too much alcohol is harmful to you, lesser amounts may be beneficial." "A glass of wine is better health-wise than a can of beer or (don't even think about this) a cocktail." Both of these assertions are incorrect, although we hear them repeatedly.

A study published in 2018 by the French Institute of Health and Medical Research in the *British Medical Journal* touted that people who drink no alcohol were more likely to develop dementia than those who drink moderately. In other words, a little bit is good for you. Before accepting that assertion, consider this. The nondrinking group included individuals who had been drinking earlier in life but discontinued it. It's quite likely that at least some participants in this group had been advised either by professionals or family members that they should decrease their alcohol intake. If so, it seems reasonable to assume that they were abusing alcohol to a greater or lesser extent.

Such "situational artifacts" as Christopher Labos, MD, an epidemiologist at Queen Elizabeth Hospital Complex in Montréal refers to them, make it more difficult in any study of nondrinkers versus drinkers, since at least some of the so-called nondrinkers are skewing the numbers secondary to the adverse effects of alcohol coming on years after they have stopped moderately heavy or heavy drinking. In order to make a claim that occasional drinking may be better for your health than no drinking at all, researchers should include in future studies within the alcohol abstinent group only those who for one reason or another (religion, ethnicity, fears of becoming an alcoholic, etc.) never drank alcohol.

The harmful effects of alcohol include certain types of cancer such as head and neck (pharynx and larynx), breast, liver, esophagus, and colorectal. Liver cancer is the most common and accounts for the greatest number of alcohol-related deaths. What's particularly pernicious about alcohol-induced liver disease is that the only way of diagnosing it in its earliest stages is by an often serendipitous comprehensive metabolic profile (measuring numerous blood values), typically taken as part of a routine physical. While the resulting liver-function test results are markedly abnormal, cluing the physician to possible alcohol abuse, the patient shows none of the symptoms that will soon follow if liver function further deteriorates secondary to the alcohol abuse: nausea, vomiting, a yellow tinge to the eyes and other areas of the body (jaundice), and abdominal pain. On testing, nine out of ten people who regularly consume more than four drinks a day display signs of alcoholic fatty liver. This should not be surprising, since alcohol is a poison that kills a certain percentage of people yearly, either by acute toxicity following an especially egregious bender, or slowly over the years secondary to a breakdown in one or more bodily organs.

Memory is especially vulnerable to the destructive effects of alcohol. Korsakoff's syndrome, marked by a profound loss of recent memory, results from the direct effect of alcohol on the brain within a setting of reduced levels of thiamine, a vitamin that helps brain cells produce energy from glucose (sugar). When the thiamine drops to a certain point, brain-cell function deteriorates, sometimes precipitously. Within an hour, a normally functioning heavy drinker may become confused, lose balance, stagger, and fall. Most affected is the memory for recent events, along with memory gaps that are filled in by confabulation—making up scenarios that seem plausible but have never happened. Here is a demonstration of confabulation from a short conversation I had with a patient suffering from Korsakoff's:

Dr. R.:   "Yesterday when I was shopping at Paul Stuart's clothing store, didn't I see you there looking at shirts?"

Patient: "Yes, I was there. I didn't see you, but I purchased a nice shirt."

Dr. R.:   "What color?"

Patient: "Blue. That's my favorite color."

Dr. R.:   "Did you find a tie that looked good with it?"

Patient: "I bought a red tie because I think red goes nicely with blue."

This man wasn't lying. He readily took up and believed my suggestion that a fictitious encounter in the clothing store actually happened. The fact that he hadn't been out of the hospital in weeks had no effect on his going along with my suggestion.

If left untreated, Korsakoff's progresses to a permanently impaired memory and finally an alcohol-related dementia. If thiamine is administered early enough, a full recovery is possible. This makes Korsakov's a reversible dementia and a true neurological emergency: If the diagnosis of a deficiency isn't made early enough, and supplemental thiamine is not provided, brain cell destruction inevitably results

In about 25 percent of Korsakoff's patients, the peripheral nervous system is affected as well—the reason for the loss of balance and frequent falls. As a consequence of memory loss, learning and reasoning difficulties soon follow. Although the Korsakoff patient's difficulties are observable to even a casual observer, the patient himself lacks all insight into his condition, frequently sees and hears things that aren't there (hallucinations), expresses increasing confabulation, and undergoes a change in personality, usually adopting an irascible, even paranoid, combative orientation toward others.

Although Korsakoff's is an extreme example of the harmfulness of alcohol, it illustrates the point I made earlier that alcohol is a direct neurotoxin, in this case secondary to the damage that it causes in the presence of low levels of thiamine, secondary to the dietary deficiency which is only too common among severe alcohol abusers. Fortunately, Korsakoff's is becoming rare, because of the fortification of foods with thiamine and other nutritional supplements.

Because memory is so important to our identity and functioning, the issue of drinking versus nondrinking takes on an oversized importance. But that isn't to deny that some people take great pleasure in the occasional glass of wine. Keep in mind that alcohol is only one risk factor for Alzheimer's disease. It is less harmful to the brain than heavy smoking, for instance. Further, risk factors must be put into context. The sum of all risk factors is important, rather than the effect of just one of them.

That said, alcohol should also be seen in the context of frequent falls among the elderly. The death rates from falls is increasing, especially among elderly men. Death rates subsequent to falls increased 30 percent between 2007 and 2016. Falls were responsible for 70 percent of accidental deaths in people of 75 and older.

That's why I strongly suggest that if you are 65 years of age or older, that you completely and permanently eliminate alcohol from your diet. You should especially heed my suggestion if you are already afflicted with other contributors to falls, such as a decline in strength, muscle atrophy, balance issues, and the taking of medications. In that case, drinking alcohol may be especially dangerous.

In order to make your own decision about alcohol in your life, ask yourself, *Why do I drink?* If the answer is "Because alcohol helps me to elevate my mood and lower my anxiety," you may be at some peril, and it's probably best to stop altogether. If your answer is something akin to "Wine enriches my enjoyment of food and the people I'm with" or "I like to sample a bottle of wine with my friends and discuss our impressions," you probably have less to fear. Hey, nobody is claiming you can't occasionally engage in some low-risk activity. Nobody is perfect. *Le mieux est l'ennemi du bien.* (The perfect is the enemy of the good.)

In addition to the foregoing commonly accepted risk factors for dementia, three other factors for preventing dementia are important.

First: Eliminate perceived stress

Although stress reduction seems a commonsense way of lengthening the odds of dementia, the key descriptive adjective is usually omitted. *Perceived* stress is the key qualifier. Perceived stress is a consequence of events or demands that exceed in individual's self-assessed ability to cope. It varies from one person to another, even when they encounter similar situations. "One person's stress is another person's challenge" sums it up. For something to be considered a stress, the person must perceive it as such.

Perceived stress also changes as circumstances change. In tennis, for example, an unranked player may defeat a highly ranked one because, basically, the lower-ranked player has nothing to lose and thus doesn't find the match stressful. But if the lower-ranked player remains superior and in control throughout the match, the higher-ranked player begins to come under perceived stress because of the threat of losing to an opponent deemed inferior. As another example, most people consider holiday gatherings relaxing, enjoyable, and nonstressful. Others perceive as stressful the prospect of spending enforced time with an in-law or even a direct relative. Among older people, the perceived stress typically ratchets up even more.

Perceived stress, according to experimental physiological and psychological research, serves as a modifiable risk factor for mild cognitive impairment (MCI) and perhaps later Alzheimer's. Among racial and ethnic minorities, perceived stress not only directly affects cognition but also contributes to the worsening of unhealthy behaviors, such as excessive drinking, smoking, physical inactivity, and inattention to and poor compliance with medical advice.

Perceived stress increases linearly with age, and it also elicits a faster rate of cognitive decline among elderly people. The mechanism involves stress hormones crossing from the blood to the brain and inducing brain atrophy and concomitant cognitive decline. These findings are from a 2023 study, "Association of Perceived Stress of Cognition among Older US Adults."

How to reduce stress among the elderly? Simple. Ask the question, "What really stresses you out?" Although the reasons a person experiences stress—often under seemingly nonthreatening situations (i.e., at least nonthreatening to the person asking the question)—may remain hidden, the experience of perceived stress is always accessible. You just have to ask.

Perceived stress, I believe, explains many people's attitudes about dementia. They perceive themselves as likely to come down with it, and the ensuing stress becomes chronic, mostly in response to the inability to *guarantee* that they won't develop it. Unfortunately, this stressful response, if pervasive, leads—in a kind of self-fulfilling prophecy—to the very result that is most feared: deterioration in cognitive function that may meet the criteria for MCI or even one of the dementias.

One of the positive aspects of perceived stress is our ability to decrease it by as something as simple as changing our attitudes and expectations.

Second: Rethink your relationship with recreational drugs

Throughout the nation, we are presently witnessing an unprecedented liberalizing of our acceptance of marijuana (cannabis). Much of this is fueled by special interests, who presciently foresaw a huge financial windfall from seizing the limited distribution channels for the legalized marijuana currently available. These same special interest groups now control the narrative about the dangers of marijuana. Neuroscientists who point to the negative consequences of marijuana have been largely ignored or subtly excluded from the conversations concerning risks.

In an attempt to avoid all this, let's start with commonly accepted observations about marijuana (commonly accepted, except perhaps by those who have an economic and/or political axe to grind). *Marijuana* is the term used for the dried flowers and leaves of the cannabis plant. The plant's main psychochemical, THC, causes its effects by attaching to THC receptors, especially those receptors that are important for memory function, like those in the hippocampus, amygdala, and cerebral cortex. The brain's concentration of THC receptors is highest in those brain areas.

On the positive side, marijuana has been shown to be sometimes helpful in relieving pain and nausea. Other claims—including one that the drug helps in the treatment of dementia—have been shown to be untrue. Although some physicians treating Alzheimer's patients observed some mild calming of agitation, the use of marijuana in anyone over sixty years of age is contraindicated. Why?

Pot has become much more potent over the years. In 1980s and 1990s the potency was 4 percent of THC, a decade ago 12 percent, and currently 26 percent. As a result of lack of knowledge about the more potent marijuana, cannabis-related emergency-room visits by heavy marijuana users over age 65 have risen by 1,800 percent among Californians, according to a study from the University of California, San Diego.

A more neutral and reflective look at marijuana use turns up certain problematic facts. Long-term cannabis users—those who started using marijuana at 18 or 19 and continued through at least their mid-40s—show cognitive defects in memory. That's unsurprising, as the brain's memory circuits possess the highest number of brain cannabinoid receptors.

A study from New Zealand found that frequent marijuana use (usually beginning in adolescence) was associated with the loss of six to eight IQ points when measured in midadulthood (forties). Cannabis use continues this trend among baby boomers (the generation born 1946 to 1964), who now make up the over 60-year-old cohort and are responsible for the 1,800 percent increase in ER visits secondary to unfavorable reactions to the increasingly potent marijuana now available. What's more, adolescents who use marijuana regularly are less likely to finish high school or obtain a degree. They also are at increased risk of using other brain-altering drugs and attempting suicide, compared to non-pot users. Although each of these latter factors forms a correlation link, a definitive cause-effect remains speculative (see our earlier discussion of correlation versus causation). Other factors associated with marijuana use may also exert an effect, as might previously demonstrated associations, including lower income, welfare dependence, unemployment, and lower life satisfaction.

What is not speculative is marijuana's effect on the brain as it affects episodic memory, working memory, short-term problems with thinking, executive function (judgment), and physical actions that at least initially depend on conscious thought, such as learning to operate a machine or play a musical instrument. Self-perceptions are also skewed in someone who is taking marijuana.

In one study, when asked how marijuana affected their thinking, career achievements, social lives, and mental and physical health, the majority of heavy users reported that marijuana had exerted negative effects on all of these five areas of their lives. Researchers have also found that psychotic illness was significantly more prevalent among those who had used marijuana. Supporting this subjective impression is a study among postal workers. Postal employees who tested positive for marijuana in their pre-employment urine drug test sustained 55 percent more injuries and a 75 percent greater absenteeism compared to those with negative marijuana pre-employment drug testing.

According to Madeline Meier, PhD, the lead investigator on the Dunedin Longitudinal Study in New Zealand, "The deficiencies we saw among long-term cannabis users . . . were in the range in terms of effect size of what we see among people in other studies, who have gone on to develop dementia in later life."

The brain effects of the currently available marijuana are much greater than that of just a few years ago when the THC level was lower. Therefore, using marijuana now puts you at a risk for brain damage and dementia. So, it would seem a no-brainer (excuse the pun) to abstain. But powerful invested interests are intent on extending marijuana use not only for the occasional therapeutic benefit but also for full time recreational use. Unfortunately, it's all part of a health-care pattern that's becoming all too common. Currently in the United States, medical issues like the use of marijuana or the development of drugs aimed at preventing, or at least delaying, Alzheimer's are not just medical issues but also cultural, economic, and political ones. More about this in Chapter X.

Third: Experience the magic of the tree canopy

If you've lived most of your life in a big city, have you ever wondered whether you might be happier in a rural environment? If you grew up in a small town (as I did), do you think you might be happier in a big city? Whatever your answers, health issues would probably play some role in your decision. In order to formulate that, you might seek out information about the roles of country versus city living in regard to cognition and the likelihood of dementia.

According to a study from Ireland (the Irish Longitudinal Study on Aging), city residents display a cognitive advantage compared to their country cousins. The Irish Study involved 3,765 healthy Irish residents aged 50 years or more. "Evidence of higher prevalence of dementia in rural, rather than urban, context suggests that urban environments may be more stimulating, either cognitively, socially, or in terms of lifestyle," the authors concluded.

While the findings and reasoning in the Irish Study seem to make sense, I'm not entirely convinced. I've lived on both sides of the rural-urban divide. The first two decades of my life I spent in a rural setting and the rest of my life, so far, in urban settings (New York City and Washington, DC). At the very least I'm wondering if that purported urban advantage is becoming increasingly tenuous in response to air pollution and global warming.

Higher concentrations of nitrogen dioxide (NO_2) and carbon monoxide (CO) are associated with greater risk of dementia. Measurements of traffic-induced air pollution, for instance, indicate that the closer you live to a major and much traveled highway, the greater your particle exposure and your chance of cognitive impairment. Long-term exposure is correlated with faster cognitive decline.

A good summary of these findings can be found in "Traffic-Related Air Pollution and Dementia Incidence in Northern Sweden," authored by the famed researcher Lars-Göran Nilsson and published in 2016 in *Environmental Health Perspectives*.

The air pollution–cognitive impairment nexus does not require long-term exposure. Just two hours of exposure to diesel exhaust can interfere with normal brain function. Measures of brain activity during that time showed decreased functional activity in widespread brain areas after exposure to diesel exhaust when compared to filtered air.

In light of these findings, I think we should take a second look at the correlation between urban or rural cognitive impairment and dementia. For instance, it is known that running and other active physical pursuits outdoors are more likely to promote mental health benefits compared to carrying out the same activity indoors under artificial conditions (running on a mechanical treadmill in a gym, for instance). Within either an urban or rural setting, the density of the tree canopy is positively related to mental health and the sense of well-being reported by the runner living nearby that tree canopy.

Unfortunately, we cannot take the tree canopy for granted. Toronto, one of my favorite cities, with its vast two million acres that form an arch around the city (the so-called Greenbelt), is currently experiencing a bitter disagreement about trees. In response to the flood of immigrants pouring into Canada and Toronto in particular, new homes are proposed to be constructed on privately owned parts of the Greenbelt. Whatever the other political-social-cultural effects of this transmogrification may be, the tree canopy may be reduced—leading to some as yet to be determined degree of pollution. No doubt Toronto should be among the cities currently aiming at discovering the optimal balance between the cognitive benefits of a city versus an urban environment.

In the meantime, until we know more, the best course of action seems to involve combining these elements: choose a medium-to-large city with all the intellectual benefits that you may not be as likely to encounter in a small town. But choose for your residence an area in that city that contains generous amounts of green space, and live close to or within that green space. I followed that advice myself. I live in Washington, DC, but in a residential area surrounded on three sides by woods.

A MAGNIFICENT OBSESSION

Let's close this chapter with the most important component in our armamentarium against dementia.

A hurricane destroys two houses. One owner is wiped out financially since all of his assets were tied up in his house. The owner of the second house shrugs off the destruction of his property, turns everything over to a caretaker, and flies in his private jet to his multimillion-dollar digs in Manhattan. An MBA isn't required to appreciate the fact that wealthy people are in a better position than their less-prosperous neighbors to withstand financial hits. From the strictly economic point of view, we could describe the wealthier man as possessing a greater financial reserve.

The brains of different people also differ. As with wealth, cognitive reserves are built up over time. *Cognitive reserve theory* refers to the representation stored within the brain of the knowledge, experience, and life events that accumulate during the course of a person's lifetime.

Education is the keystone of cognitive reserve. The more education a person achieves, the greater the cognitive reserve with *one critical qualifier*. Education isn't just the accumulation of degrees and diplomas. Even more important is the education that occurs after the diploma has been placed on the wall.

As an illustration of this point, consider a man I met on an Egyptian educational trip about twenty years ago. A typical day involved touring various pyramids and other sights, to the accompaniment of the commentaries given by our Egyptologist guide. At the end of the day, we would all meet for a Q-and-A session with the guide. One of our group, let's call him Frank, seemed to know much more than anyone else, with the exception of the guide.

While the exchanges between the rest of us and the tour guide took the form of ask-the-expert exercises, Frank's interactions were different. It turned out he was quite an expert himself on Egyptian history and culture. His interaction with the guide took the form of ongoing and sometimes complicated discussions between two experts. On occasion, the guide, after replying to a question from one of the other members of the group, even asked Frank if he had anything to add.

Curious, I went out of my way to learn more about Frank. While most of the tour members were the usual mix of lawyers, doctors, and entrepreneurs of various sorts, Frank, as he put it, was "strictly blue collar" and ran a successful contracting business. "So when and how did you learn so much about Egyptian history?" I asked him.

Frank still remembered the lesson about ancient Egypt his teacher gave to his fifth grade school class. This ignited a lifetime effort to learn more and more. Over the next forty years he collected and read numerous books on Egypt, attended lectures on various aspects of Egyptian culture, and joined learning tours about Egypt, such as the one we were then participating in.

In essence, Frank had developed what I call a *magnificent obsession*: an intense interest in a specific subject accompanied by long-term efforts to learn everything possible about it. Best of all, the development of a magnificent obsession doesn't have to start in childhood; it can be taken up at any age.

Another way of thinking about cognitive reserve is to consider it as a part of crystalized intelligence—the factual knowledge that accumulates over the years from education and experience. This does not noticeably decrease with age and may even improve. When you consult with an elder expert about a particular course of action you should take (in legal, medical, or other professional matters), you are drawing on the elder's accumulated crystalized knowledge, sometimes informally referred to as "wisdom."

In a study with 1,184 participants, the MRC National Survey of Health and Development (1946 British Health Cohort) that lasted for seventy years, the results showed the rate of decline in cognition from childhood to adulthood was greater in those with lower cognitive reserve and less crystalized intelligence. The findings provide strong support for the claim that a lifetime investment in building up cognitive reserve leads to healthy cognition and thinking later in life.

Of course, nobody can wave a wand and return a person to a time seventy years ago. So the issue arises whether there comes a time when it's too late for cognitive reserve to make a difference. Can cognitive reserve increase in people in their fifties, sixties, and older?

The most startling and original insight about the brain might be *plasticity*. Thanks to plasticity, the brain's functional capacities hold steady and may even increase (due to crystalized intelligence), even while the number of brain cells decreases. What's more, this remolding of the brain's circuitry via plasticity takes place in real time, measured in microseconds. This is far different from any mechanical device. Imagine opening the hood of your car, removing at random several parts, and then expecting your car to run *more* efficiently.

Whenever we undertake new interest or activities, we form new networks within the brain; the numbers of the brain's chemical messengers (neurotransmitters and the receptors for these neurotransmitters) increase or decrease according to demands. Nerve cells even have the potential to start making different neurotransmitters. Dopamine-producing neurons for instance, may start making and releasing alternative neurotransmitters. The end result of these modifications is a fundamentally different brain.

Initially, the concept of the brain's lifetime capacity for change is often difficult for many people to accept. This reluctance is based on the fact that the brain's alterations aren't subjectively appreciated until late in the process. For instance, if you start taking a French-language class, your initial learning will be slow and incremental. But over the span of a year or so, your facility with French will improve until you will hardly remember your initial stumbling efforts. But facility in the new language won't occur overnight; you won't suddenly wake up one morning speaking perfect French. That's because a certain amount of time and effort is required to establish the brain circuitry dedicated to that second language. The process of brain modification accompanying new learning, although subtler, isn't that different than from what happens in an exercise program.

It is never too late to build your cognitive reserve. It will not be done in a day or week or month. But no matter how old you are when you start, the process will take its natural course depending on your age. Building up cognitive reserve can and should be done at any age.

The easiest way of doing this? Pick something that really interests you, grips you in a visceral sort of way, and then obsess (in a good way) about it. Something like Frank's Egyptology may be what you pursue, but the subject of your obsession doesn't have to be academic or scholarly. It can be anything, even topics like becoming a superior chef. To make this happen, read recipes and cookbooks, watch videos, interview anyone who knows more about the subject than you and is willing to talk with you.

The brain remains highly malleable throughout the lifespan, and cognitive reserve can be built up starting in childhood or any time during the next seventy years, as illustrated by the MRC National Survey of Health and Development.

Take comfort in the the brain's malleability. As you read these sentences and learn new information, your brain circuitry is undergoing modification. The extent of this modification is determined by your previous life experiences. If you learn a lot, the brain circuitry changes extensively. Learn only a little bit, and your brain's organization remains about the same. Cognitive reserve is the most important concept to keep in mind in your efforts to prevent or stave off Alzheimer's.

Reading for pleasure is perhaps the single most effective activity you can engage in for increasing cognitive reserve. The practice exerts its most powerful effect on memory, both episodic and working. Episodic memory, as you recall, is concerned with the specific recall of characters or events from earlier in the book. Working memory is the capacity to hold in memory a character or series of developments throughout the length of the book.

Fiction is more brain-challenging than nonfiction, because it demands your full awareness of the narrative you are currently reading, while at the same time requiring you to hold in mind the situations and characters described earlier in the book. To do this, you have to exercise both your episodic and your working memory. It's the smooth interaction of those two memory processes that enable you to keep track of what happened, perhaps a hundred pages or more earlier in the book. In a work of nonfiction, in contrast, you can usually skip around a bit, reading chapters in isolation and out of order. The nonfiction book you are now reading is deliberately constructed to encourage some latitude in the order that chapters are read.

Reading for pleasure also demands concentration, focus, and imagination (coming up with possible explanations why the characters are behaving as they are).

Does reading help to improve memory, or is the person with a strong working memory drawn toward reading, which thereby improves reading-comprehension skills? With this question, we reencounter the cause-effect conundrum. In order to explore this intriguing question, researchers at the Beckman Institute for Advanced Science and Technology designed an interesting experiment.

In the experiment, half of the participants aged 60 to 79 were randomly assigned to an eight-week program of leisure reading on their iPads (n = 38 participants). The other half worked on completing word puzzles on their iPad (n = 38). Both activities obviously call for concentration and cognitive exertion. The final results showed that the readers outperformed the word puzzlers when tested for cognitive functioning after eight weeks. This improvement was in working and episodic memory, as well as other verbal and reading skills.

When comparing the readers verses the puzzlers, "We controlled as much as we could between the activities except for the 'magic juice,' which is getting immersed in a story," according to Liz Stine-Morrow, the study's senior investigator.

I think the bottom line here is the increased benefit that results from reading books that appeal to you and can give you pleasure as compared to working (and I use the word *working* selectively) at mental challenges that may have no appeal or interest for you. "There's more promise in engaging fully in the stimulating things that we already do in our lives. That's probably the best pathway to increasing cognitive reserve and offsetting the effects of Alzheimer's disease," says Stine-Morrow.

In this section of the book, I presented multiple suggestions how to lower the odds of cognitive impairment. But there are no guarantees. Rather than allowing this absence of certainty to further deepen our anxiety, it's better if we consider this lack of a 100 percent guarantee as one component of life's most basic tenet: there are limits on our ability to foresee our fates.

If we adopt each of the elements enumerated in this chapter, we will greatly lower the odds of dementia. Our chances will improve the most if we follow all of them. As more is learned about the specific dementias, especially Alzheimer's, other elements will no doubt be added to that checklist. The tricky component to all this is that we do not understand the basis for Alzheimer's and several of the other dementias.

BE MODERATE, EVEN ABOUT MODERATION

If you adopt all the lifestyle factors discussed here, you can delay or prevent dementia. Of all the healthy lifestyle contributions, diet, intellectual stimulation, exercise, and social contact have the strongest associations with good memory. Dementia is very rare, if not nonexistent, when the memory is unimpaired.

According to one of the authors of a study comprising 29,000 adults with a median age of 72, "Although each lifestyle factor contributes differentially to slowing memory decline, a result showed that participants who maintain more healthy lifestyle factors had a significantly slower memory decline, than those with fewer healthy lifestyle factors."

As I've emphasized throughout this book, cognitive health depends on a combination of factors, rather than any one single factor. You are not going to maintain cognitive health only by eating certain foods or by only doing a certain amount of exercise. Even if you carry the APOE allele, which is the strongest known riskfactor for Alzheimer's, you can maintain good cognitive function and lessen the chances of dementia by practicing the discussed lifestyle guidelines.

CHAPTER X

THE CULTURAL, SOCIAL, AND POLITICAL ASPECTS OF ALZHEIMER'S

CUSTOMER, CONSUMER, BUT NOT PATIENT

Earlier in this book I mentioned some of the attitudes toward dementia within earlier cultural settings: an imbalance of humors, madness, etc. But our current culture is not immune from distortions in how we conceptualize dementia. Although our distortions are subtler, they are no less pernicious and harmful.

As our society has become more utilitarian, the "virtues" of efficiency and productivity have replaced humanistic motivations of care. In the new model, dementia is increasingly looked upon as a human failure of the mind for which the dementia patient bears some degree of responsibility. He or she did not eat the right things, exercise enough, keep sufficiently mentally stimulated, etc. Beginning sometime in the early 1980s, our cultural terminology has shifted from medical-speak to business-speak to economics-speak.

I vividly remember a medical staff dinner meeting at an academic teaching hospital in about 1982 or 1983. It included several tables of recently graduated MBAs, who represented the business group that had recently purchased the hospital. When one of them began his scheduled talk, he mentioned, among other things, "From now on we want you doctors to think of the people occupying these beds as customers, not patients. Just like any customer, they want to be pleased with the product." Although most of the doctors were caught off guard and didn't know what to make of this strange patient-to-customer transformation, the end result has subsequently become only too clear.

While most doctors continue to speak of "patients," the business model has had greater luck in transforming other aspects of the medical culture of the hospital. The patient has transformed not only to customer but to consumer as well (presumably a consumer of medical care). As a result, it is not unusual for a doctor to be asked by a nurse or nurse's assistant to "check on the *consumer* in bed 105."

Many other expressions of the business model have wended their way into the hospital. The meeting of a patient with his or her doctor has mutated from an "appointment" to an "encounter"— the word *encounter* appears in hospital charts, insurance forms, and even some doctor's and nurse's notes. Added to this are the endless meetings followed by encounters, during which doctors make little eye contact while typing into computers and simultaneously talking to their patients—all in the name of increased efficiency.

To put these developments into context, 1981 was the year Harvard professor Derek Bok wrote in *Harvard Magazine*,

> The University's reliance on the industry funding for research was causing an uneasy sense that programs to exploit [i.e., make money from] technological developments are likely to confuse the University's central commitment to the pursuit of knowledge.

When Bok wrote these words, more than 80 percent of pharmaceutical research was being conducted in university medical centers and published by independent academic researchers. By 2004 only 26 percent of such research was being conducted in universities. Overall research was moving swiftly from universities to the pharmaceutical industry. Today the percentage of research done by the pharmaceutical industry is close to 100 percent, with the drug companies using and controlling all of the resulting findings and data.

But the biggest takeaway about business appropriating health care comes from efforts involving some seemingly unlikely corporations. On July 22, 2022, Amazon announced its plan to acquire a primary care practice (One Medical) with nearly 200 locations serving an excess of 700,000 patients. This $3.9 billion acquisition was approved on February 22, 2023. "Transforming Primary Care" is the logo that promises "seamless access to comprehensive care and calming offices."

The impact of this acquisition looks to be big. For patients, primary-care serves as the front door for health care in general. For investors, primary care also serves as a front door for revenue streams comprising hundreds of thousands of patients. This includes the patient's medical files and data, which are potential sources for additional revenue income based on patients serving as potential "customers" for other services, many of which have nothing to do with medicine. The Amazon purchase is not a one-off. Between 2010 and 2021 the total capital raised for private investment in primary care rose by a factor of more than a thousand—from $15 million to $16 billion.

Corporate-owned primary-care practices (CPCPs) break down into three categories, none of which traditionally plays any part in medical care: retail-owned (Amazon and Walmart), insurance-owned (United Health Optum, Humana), and investor-backed.

Commenting on this corporate-investor capture of the primary-care market, Soleil Shah writes in the *New England Journal of Medicine*, "CPCPs whose duties to maximize profits for shareholders and investors may conflict with what's best for patient care." I'll have more to say in a moment about the effects of corporate and investor involvement in the development of agents capable of delaying or preventing the onset of Alzheimer's and other dementias.

HOW CLOSE ARE WE TO A CURE?

A cure for a disease is almost always preceded by an understanding of its cause and its mechanisms of expressions. Although there are exceptions to this rule, they are few and far between. The lack of a satisfactory explanation of the cause or causes of Alzheimer's underlie our incapacity to control it.

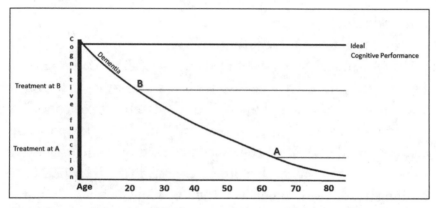

The above graph contrasts ideal cognitive performance with dementia. If dementia is identified early enough (treatment at B) considerable recovery can be expected after the development of a successful treatment. If the identification comes later, even powerful antidotes will be less effective because considerable brain damage will have already occurred, and as a result recovery will be limited.

Today, more than a hundred years after Alois Alzheimer demonstrated amyloid plaques and neurofibrillary tangles in the brain of those patients he diagnosed with that eponymous disease, the lions' share of research has been devoted to one underlying belief: the plaques and tangles are the *cause* of the disease. Yet this entrenched belief has spawned little progress.

Clinical trial after clinical trial (at least seventeen known trials) aimed at curing Alzheimer's by lowering amyloid has failed. As one expert phrased it, anti-amyloid drug development is "grounded in a mountain of biological evidence so vast and diverse, that it seems unimpeachable. On the other hand, the track record of drug after drug, trial after trial, year after year is so long that it has seems to have reached the point of sheer folly!"

Despite amyloid plaques serving as a defining feature of Alzheimer's and drug trials aimed at eliminating amyloid plaques, no drug has been developed that reduces amyloid and at the same time satisfactorily improves mental functioning in humans. Especially interesting, a few of the string of failed drugs had just about cleared the amyloid from the brain but still failed to bring about satisfying clinical improvement in cognition.

SCIENTIFIC INQUIRY OR HEDGED BET?

Alternatives to the amyloid theory are already in play. Take the research of Rusty Gage, president of the Salk Institute for Biological Studies at La Jolla, where he is the lead researcher in age-related neurodegenerative diseases. His research is focused on individual neurons taken from the brains of people with Alzheimer's. He and his team have found that many of these neurons deteriorate and undergo life stress responses called *senescence*.

Senescence can be a normal process with age elsewhere in the body, such as skin wrinkles and deteriorating eyesight and hearing. But senescence in the brain is always abnormal. As Gage and his team have discovered, senescence causes neurons to lose functional activity and express impaired metabolism, along with increasing brain inflammation. Gage's team has already formulated ways of determining the deterioration process, so preventing neuronal deterioration could be an effective approach to improving or curing Alzheimer's.

"Our study clearly demonstrates that these neurons are going through the deterioration process senescence and that is directly related to neuro inflammation and Alzheimer's disease," according to Professor Gage.

As neurons deteriorate, they release inflammatory agents that set off a cascade of inflammation traveling from one neuron to another. This can be significant since a single neuron can link with a thousand or more other neurons. The aim of this alternative approach involves targeting the senescence neurons in order to slow the inflammation and the consequent whole brain degeneration of Alzheimer's disease.

If one wants to explain the role of inflammation in Alzheimer's, allergic diseases would be a good place to start. Research has revealed that allergic rhinitis, asthma, and atopic dermatitis (a group of skin conditions marked by red itchy lesions, essentially eczema) carry an increased risk of dementia.

In a Korean study from the Korean National Health Insurance System comprising 6,785,948 people, researchers found that each of these three allergic diseases was associated with a significantly increased risk of dementia, especially Alzheimer's and vascular dementia. Is the mechanism based on cell activation in the brain producing inflammation, resulting in dementia?

But if you suffer from any of these allergic diseases (as do some 30 percent of the population), panic isn't warranted. The risk on an individual level of asthma, allergic rhinitis, or atopic dermatitis leading to dementia is small, on the order of a 10 to 20 percent increase in dementia risk.

But if follow-up research supports this link between allergy and dementia (the almost seven million people in the Korean study is by far the biggest statistical evaluation ever carried out), the implications for further research on an inflammation-modifying drug for Alzheimer's would be urgently warranted.

Instead of putting all the eggs in one basket, so to speak, why not develop drugs that inhibit the inflammatory response itself? As Rusty Gage's work suggests, inflammation may well be the initiating event leading to the death of neurons, followed by their aggregation into the amyloid bodies and neurofibrillary tangles that characterize Alzheimer's disease.

This senescence theory remains—along with the amyloid theory—unproven, with a lot research still needed to resolve such questions as "What is the benefit of removing these senescent neurons? How do these senescence neurons, left unchecked, lead to Alzheimer's disease?"

So which theory about the origin of Alzheimer's is correct? At this point no one has the answer. Should research continue to be focused on amyloid elimination, or should more firepower be directed to inflammation? Or is a combination of causes something to be expected?

So far, the predominant theory—beta amyloid—has held sway by concentrating on the amyloid plaques and neurofibrillary tangles found in the brain of Alzheimer's patients. This theory rests on the assumption that Alzheimer's can be understood by considering how the plaques and tangles led to Alzheimer's. Examining and probing into the nature of amyloid and neurofibrils may enable a neuroscientist to reverse engineer what leads to the formation of those amyloid bodies and neurofibrillary tangles. The process is much like a four-star chef tasting a dish and divining the recipe and procedures needed to duplicate it.

But perhaps the true cause involves a yet-to-be-determined injury by an unknown cause, followed by a powerful cascade of inflammatory processes. In this scenario, the cause will emerge from concentrating on inflammation. The amyloid plaques and tangles are merely the end products of the inflammatory cascade or something currently unknown that destroyed neurons and, as a side effect, produced the plaques and tangles that on their own are unlikely to provide the key to understanding Alzheimer's. At least they haven't so far.

So which of the two most discussed theories—amyloid and inflammation—is correct? Are both of them wrong? Are both of them complimentary and thus both partially right? Perhaps the search for a single cause is misguided. An analogy by Dr. Donald Weaver, senior scientist at the University Health Network's Krembil Brain Institute in Toronto, compares the situation to treating high blood pressure.

"There isn't one pill for high blood pressure," according to Weaver. "So why do we expect there is going to be one magic bullet, one pill, that's going to be the cure for Alzheimer's disease? I think that's naïve."

THE BATTLE OF PHARMA TITANS

In 2020 the pharmaceutical companies Biogen and Eisai were seeking approval from the FDA for an anti-Alzheimer drug called aducanumab (brand name Aduhelm). During the approval process, the FDA ignored the advice of its own Advisory Committee. Not one member of their select committee of experts on Alzheimer's advised approval of the drug. The FDA took an unusually trenchant position by accelerating the approval process. A lot was at stake. Biogen estimated that the drug could bring in $18 billion a year. In one slide presentation, the company boasted, "Our ambition is to make history and establish the drug as one of the top pharmaceutical launches of all time."

Initially, the cost of Aduhelm for patients was $56,000 a year, which when spread across the Alzheimer's population in the United States would come to $12 billion the first year. While such a windfall would be great for Biogen's investors, half of the potential users of the drug earned incomes less than $50,000 a year. In response to insurance companies' refusal to cover the drug, Aduhelm was rarely prescribed, and at the end of the first year, it was responsible for earnings of a mere $3 million. At this point the plug was pulled on the drug.

Within a year of the discontinuation of Aduhelm, another drug, lecanemab (brand name Leqembi), once again from Biogen and Eisai, was introduced. Showing only low-to-moderate benefit, the drug during clinical trials was associated with increases in brain swelling and brain bleeding when compared to patients on placebo. Putting the best face on it, the benefits were hardly overwhelming. As one researcher, Dr. Madhav Thambisetty, a neurologist and senior investigator of the National Institute on Aging, said, "From the perspective of a physician caring for Alzheimer's patients, the difference between Lacanemab and placebo is well below what is considered a clinically meaningful to treatment effect."

So what was the rush to approve a drug with such a narrow window of effectiveness? Repeated development and testing of anti-amyloid drugs might be secondary to what Harvard Medical School and world-acknowledged memory expert Steven Schacter refer to as one of the "seven sins of memory": persistence. It seems that once an idea becomes engrained in our brain, we have great difficulty replacing it or even modifying it. This reminds me of an ersatz definition of insanity: "doing the same thing over and over under the same circumstances and expecting each time to get a different result."

After many failures producing an amyloid-based treatment for Alzheimer's, it becomes easier to start yet another amyloid trial rather than considering the possibility of some other cause of the disease. This is not to say that the amyloid hypothesis isn't correct—there are dozens of research teams throughout the world headed by brilliant neuroscientists who are highly convinced that the amyloid hypothesis is the path that should be followed. Nevertheless, equally brilliant neuroscientists like Rusty Gage feel that other paths have been insufficiently explored.

In early January 2023 the FDA approved lecanemab for the treatment of early-stage Alzheimer's. The lion's share of this patient group (85 percent) was insured with Medicare. In a catch-22 consequence, the vast majority of eligible Alzheimer's patients would not be able to take the drug under these circumstances. Why? Because Medicare won't pay for any drug unless the patients are first enrolled in a government-approved clinical trial. Lecanemab was never subjected to such a clinical trial.

In the Spring of 2023, Eisai and Biogen conducted an extension study incorporating findings on lecanemab since its release (a so-called post-marketing trial). Based on this study, all six members of the FDA advisory committee voted in support of the effectiveness of lecanemab. Some questions, however remain concerning: Three people taking lecanemab died; brain atrophy accelerated compared to patients not on lecanemab. In all, a higher risk was found among APOE4 homozygotes (two genes from APOE4) and among people with amyloid within the cerebral blood vessels, as well as people who require ongoing treatment with anticoagulant agents (blood thinners). Despite these relatively infrequent side effects, it's my belief that lecanemab will go on to receive a full approval. This will mark the first time in twenty years that the FDA has bestowed full approval on an Alzheimer drug. Most important, this will enable millions of Medicare recipients to be eligible for the drug.

As an unforeseen consequence of the post-marketing findings, it will be necessary to carry out APOE4 testing to gauge risks in an individual patient. In the presence of two genes (homozygous for APOE4), a patient is six times more likely than an APOE4-negative patient to experience such side effects as brain swelling or effusions, and three times more likely to suffer a series of micro hemorrhages or even the more serious macro hemorrhages (brain bleeds). Such findings dictate that potential patients for lecanemab treatment undergo APOE testing prior to treatment. So serious are the potential consequences that in February 2023 the Department of Veteran Affairs Pharmacy Service and Medical Advisory Panel included APOE4 homozygous (two copies) on a list of exclusion criteria.

As one final complication, drug's approval for early-stage Alzheimer's is oftentimes a treatment in search of a disease. As discussed earlier, early-stage Alzheimer's is difficult to determine and confidently distinguish from MCI or other dementing illness, regardless of amyloid plaques.

THE *SHARK TANK* EFFECT

The operating principle behind all this delay in developing a successful medication for Alzheimer's, I believe, is based on what I call the "*Shark Tank* effect." If you have never seen that television program, a small group of multimillionaires audition ambitious entrepreneurs who lack the funds to bring to market a novel idea that could potentially be worth a lot of money. The goal of the multimillionaires is to acquire the greatest share of a potential future money earner, while at the same putting out as little seed money as possible. The entrepreneur, on the other hand, wants to get as much money as he or she needs to get started without giving away too big a financial interest in the future company, which will be formed by the entrepreneur and the "winning" multimillionaire who offers the sweetest proposal.

Unfortunately, the *Shark Tank* factor is not limited to television shows. An unprecedented number of billionaires and millionaires in this country are in fierce combat with each other to control formerly open market forces. A striking example of this process is the rental housing market, which has now become prohibitively expensive as a result of real estate trusts increasing their share of apartments from 44 percent in 2011 to 70 percent in 2022, according to data and research from MSCI (Morgan Stanley Capital International).

What's the difference between conventional rentals and renting from a real estate trust? While apartment companies typically deal directly with their customers (the renters), real estate investors and firms aim their efforts at pleasing their customers, the investors in the real estate trusts. What's good news for the investors—increase in the cost of rentals—is very bad news for the renters.

In a similar way the *Shark Tank* factor has become the propulsive force in product development of new drugs. Most of the world's drug manufacturers are publicly traded—that is, you and I can purchase shares in the company. And indeed, as with every other publicly traded company, pharmaceutical companies, if they want to survive, are under relentless pressure to please their investors, who like investors in any company are expecting a huge return on their investment.

In response to the financial loss due to Aduhelm, Biogen in February 2022 announced plans to lower its expenses by half a billion dollars. Further restrictions were planned, if the company's outlook didn't improve. This cost-cutting effort was needed because the drug had performed so poorly in the market, most likely in response to the lukewarm reception it received from insurance and FDA statisticians, who believed there was insufficient evidence of effectiveness to approve the drug. When the sales figures proved disappointing secondary to tight regulations by the FDA, Biogen had no choice but to withdraw the drug from the marketplace.

What was the response of Biogen shareholders to all of this? They filed a lawsuit in February 2022 against Biogen. Investors alleged Biogen's contacts with the FDA were unlawful and that the company made twenty-five false and misleading statements about Aduhelm. The suit, filed in Massachusetts, was dismissed in March 2023. "A securities fraud complaint cannot rest on a house of cards made of mischaracterized statements," wrote the judge in the case.

While a *Shark Tank* arrangement may be fine for many entrepreneurial ventures, is it really the best way of bringing new Alzheimer drugs to the market? For instance, two months before the clinical trial of lecanemab was completed, Eisai and Biogen released press releases touting high-level results that, unsurprisingly, led to a dramatic rise in the companies' stock prices. Clearly with six million potential customers out there, gaining approval for lecanemab promises to turn into a financial bonanza. That's despite what Samuel Gandy, professor of neurology and psychiatry at New York's Mount Sinai Hospital said about the drug: "No one yet knows that the benefits will be clinically meaningful. . . . This may take years to determine and the debate is likely to continue."

Obviously, I'm not making any claims of malfeasance on the part of the FDA, Biogen, Eisai, or any other pharmaceutical manufacturer. Nor am I criticizing investors who try to profit by placing their bets on the winning pharmaceutical company and drug. My point is this: in the current research environment, the development of an effective Alzheimer medication will be affected by the same market forces governing any product that promises to make a lot of money for the investors in the pharmaceutical company that synthesizes the world's first successful drug. You may find this whole shabby process disappointing and cynicism-inducing, as do I, but drug development, it turns out, doesn't depend on white-coated researchers riding white horses; it is just another market-driven hustle by entrepreneurs for whom the first and only consideration is personal financial gain.

So what will happen when, despite all this, a truly effective drug becomes available? Will a *Shark Tank* mentality and modus operandi come up with a way to prevail then too?

EPILOGUE

IMPERATIVE: WE MUST CHANGE OUR CURRENT ATTITUDE TOWARD PEOPLE WITH ALZHEIMER'S

One of the questions we've been grappling with throughout this book is the continuous versus noncontinuous nature of Alzheimer's and the other dementias. Do people with dementia represent a sharp break from what's considered normal—a kind of impenetrable wall sharply demarcating people with dementia? Or is dementia—at least in its early stages—marked by a slow, sometimes difficult to discern, relentless progression? I expect that by this point in the book you know which of these two models I favor.

The Continuum of Dementia

Cognitively Normal · Dementia

But believing in the continuum model carries with it certain obligations, I believe. We really can't be certain how long the Alzheimer's person is capable of meaningful communication with others. Over the last two decades or so, neuroscientists have revised their ideas about whether seemingly nonresponsive persons retain their ability to interact with people and events in their environment. Over the last decade, brain-imaging studies have demonstrated that even a person in a deep coma often perceives and responds to the words spoken to them.

So it seems reasonable to assume that a person with Alzheimer's or one of the other dementias may be aware of comments or remarks that are said about them by others. Yet I have witnessed family members—even doctors—speaking in front of an Alzheimer patient about the incurability of the disease.

It's best to think of Alzheimer patients as occupying not one continuum, but many. In fact, each of the symptoms and signs of the disease progresses at its own pace along a continuum from slightly abnormal to grievously impaired. Thus, for example, it's possible for speech expression to be defective (expressive aphasia) while spoken speech by others may be recognized quite well. Keep such distinctions in mind as we explore the ways people with Alzheimer's are thought about, spoken to, and generally treated.

PINK PAJAMAS AND BEIGE HEELS

Our attitudes toward people with dementia are especially affected by the language and expressions we use. If we think of culture as a learning system comprised of shared ideas, preconceptions, meanings, and behaviors, we can learn some surprising things from such cultural expressions as newspapers, social media, movies, television shows, and books.

According to a study of movies by Lee-Fay Low and Farah Purwaningrum of the University of Sidney in Australia, published in *BMC Geriatrics* in 2020, the usual depiction of Alzheimer's patients employs harmful, even destructive stereotypes: people with dementia are inevitably old, seriously impaired, incapable of meaningful interaction with others, and possessing no quality of life. As a result, too many of us have assimilated these stereotypes into our beliefs.

People with dementia are routinely depicted within the "Living Dead frame," according to Low and Purwaningrum. Descriptions included "death that leaves the body behind" and "withered shells." Of twenty-three fictional movies with a dementia theme, thirteen concentrated on severe memory problems and ten on word-finding difficulties. Typically, the affected person was confused and disoriented to time, place, and the people in the immediate surroundings. Most extreme are episodes illustrating bizarre behaviors: walking into an office in pink pajamas and beige heels; taking a hammer or screw driver to a front door in response to temporarily being unable to find the key. In almost all of the films, the narrative culminates in institutionalization and death. Further, many of the characters with Alzheimer's express an eagerness to die. Those caring for them are inevitably presented as burned out and eventually in bereavement.

Since media depictions both reflect and influence a culture's attitudes and behaviors, it should come as no surprise that many of the attitudes and behaviors depicted have slowly been incorporated into our assumptions about Alzheimer's. As Low and Purwaningrum summarized their findings, "The public is afraid of dementia, our health professionals treat people with dementia with less respect after diagnosis, and people with dementia experience othering [treating a person as intrinsically alien]."

In an attempt to counteract these harmful cultural stereotypes, language guidelines were adopted in Australia in 2018. According to these guidelines, words like *suffer* and *victim* along with the use of condescending expressions like "old boy" and "gal" are to be avoided; so are movie and television depictions of only advanced cases.

With some compassion and knowledge of the disease, late or advanced cases should be replaced by depictions of milder or earlier cases. It is useful to remember that dementia, like cancer—to which it is often compared ("mind cancer" or "brain cancer")— exists in mild, moderate, and advanced forms. Often, work and professional careers can be continued for years after the diagnosis is first made, albeit with appropriate oversight. On rare occasions, for reasons that cannot currently be explained, the dementing process may even plateau at a point where adequate cognitive function still remains.

THE TRAGIC DUET: DENIAL AND STIGMA

Only one in four people with Alzheimer's have been diagnosed, according to Alzheimer's Disease International. This statistic raises an immediate question. Why so low a rate, considering Alzheimer's is perhaps the most discussed disease in the world? One would expect doctors, whatever their specialty, to diagnose it more frequently. A medical degree isn't even necessary to detect the late-stage problems with memory, orientation, and the other presentations of the disease.

Denial, of course, plays a role. It's understandable that relatives may ignore or rationalize the earliest signs of the disease in their spouses or relatives. *Dementia blindness* is the term for denial coined by Dasha Kiper, who has devoted her career to supporting and training people who care for Alzheimer's patients. She speaks of dementia blindness as the inability to accept or even recognize that someone familiar is descending deeper and deeper into dementia, despite a plethora of evidence supporting it.

More important than denial, I believe, is stigma: negative stereotyped beliefs, feelings, or behaviors directed to anyone diagnosed or even suspected of Alzheimer's ("Not in my family!"). The cumulative effect of stigma is rejection and exclusion, both fueled by an existential dread.

Stigma includes public fear of dementia, hopelessness, assertions that nothing can be done to help, the placing of dementia units under lock and key; etc. Our culture encourages an us-versus-them orientation. Alzheimer's patients possess "diseased" brains that are totally different from "normal" brains.

When a person is diagnosed with Alzheimer's, other people view them as somehow impaired, even when they are in the earlier stages of the disease; they are considered unable to make decisions for themselves; they possess no quality of life. Relatives often don't want to investigate the possibility of Alzheimer's; those experiencing the symptoms and expressing the signs of the illness deny it; culture as a whole demeans, ridicules, and conceals it.

In their study of media depictions of Alzheimer's, researchers Low and Purwaningrum found that stigmatizing stereotypes about Alzheimer's often merged with stigmatizing stereotypes about aging in general: "Older people do not contribute and are a burden to our society" was a typical expression of such ageism. This was often coupled with a mental health stigma that included such stereotypes as violent behavior.

Given all these factors, is it at all remarkable that only one in four people with Alzheimer's have been diagnosed?

WERE THE LILLIPUTIANS THE OFFSPRING OF DEMENTIA?

While most diseases have been chronicled by their sufferers, not so with Alzheimer's or the other dementias. Third-person accounts of Alzheimer's are plentiful, if you restrict your search to narratives composed by relatives or spouses. The biography of novelist Iris Murdoch by her husband, literary critic John Bayley, stands out. As far as I know, Murdoch, one of the most prolific of Irish writers in the twentieth century, did not produce any literary product describing the encroachment of Alzheimer's into her life.

In most cases, the best that can be expected from a writer afflicted with Alzheimer's is the production of a book (usually a novel) that reflects his or her inner experience. Take the eighteenth-century writer Jonathan Swift. In his old age, Swift may have been afflicted with dementia, based on his complaints of poor memory, short temper, and a prolonged sense of hopelessness. But this is only a guess. Could it be that the tiny Lilliputians described in *Gulliver's Travels* were based on the microhallucinations that frequently accompany Lewy Body dementia? Again, we don't know.

But in terms of vividness of the self-experience of Alzheimer's, one book stands out. If you want to step into the shoes of a person with Alzheimer's, I highly recommend the autobiography of Thomas DeBaggio, *Losing My Mind: An Intimate Look at Life with Alzheimer's*. Now deceased, DeBaggio was a former journalist who recognized within himself the inroads of Alzheimer's. He was especially apt at describing the effects of Alzheimer's on his memory, identity, and sense of isolation and describing the transience of his life:

DeBaggio on Identity: "Without memory you lose the idea of who you are. I'm struggling more than ever to find answers to questions of identity. I'm flooded with early memories preserved in protected places of my brain, where Alzheimer's does not reign supreme. The memories become the last remnants of my search for who I am."

Isolation: "Every day is new now, with little remembrance of the day before, but enough memory retained to know there was a yesterday. This is a new way to live and takes getting used to."

Transience: "Memory is in the present, one minute at a time, and that is disappearing at an alarming rate. I'm truly living in the present. My memory is obliterated as I lived it. There is hardly a light to illuminate the long tunnel of yesterday. Most of my memory is obliterated the instant it is created."

SEVEN HOT-BUTTON ISSUES

In the interval between now and when a successful medication for Alzheimer's becomes available, what attitude should we take about people with Alzheimer's? It seems the better part of wisdom (and compassion) to consider the Alzheimer person as not really that much different from you. Pervasive absence of that approach when interacting with someone with Alzheimer's explains the stigmatization, the avoidance, the rising irritation, the breakthrough of anger and impatience and fear.

Conversely, many of the difficulties experienced when interacting with an Alzheimer's patient is the realization that practically all of the signs and symptoms of the illness have been experienced at one time or another by all of us: the failure to remember a name, to come up with a specific word, to recognize a familiar person when encountered in unfamiliar settings. Seeing ourselves in the patient also helps to explain the stigma and isolation. So, what to do?

Consider for a moment what a person loses when he or she is diagnosed with dementia. By keeping these seven issues in mind, great progress can be made in improving quality of life:

1. The initial most damaging loss is **loss of identity**. Included here are occupational identity, familial identity, and social identity resulting from friends drifting away secondary to their fears that something similar may happen to them.
2. The loss of the sense of **agency** means the inability to be one's boss, so to speak, and to make one's own decisions about housing and everyday activities. "Want to call up a friend and go to lunch?" Sadly, someone with Alzheimer's is rarely in a position to do that.

3. Related to the first two is the loss of the feeling of **autonomy**. The person with Alzheimer's no longer feels in control of his or her life.
4. **Always feeling alone** is a loss.
5. Always **being under critical observation** is the loss of autonomy.
6. Memory loss contributes to the loss of **familiarity** with the people and things in the immediate social environment.
7. Finally, the **loss of freedom** comes from the onerous weight imposed by boredom and repetition.

The current panacea for addressing these hot-button issues is the much touted assisted-living arrangement, which involves a person with dementia moving permanently from his or her home to a care facility.

In 2020, 818,000 people were living in assisted-living facilities for dementia, according to government estimates. Between 2015 and 2022, the number of these facilities expanded by 24 percent. Moreover, costs of these memory units, as the dementia-care units are called, rose to a price tag north of $65,000 a year, well beyond most people's ability to pay. If a locked unit is deemed necessary, the cost increases to $80,000 a year on average. As a result, the only recourse for many people is home care. But the number of home-care aides is far below what's needed. As a consequence, care often has to be undertaken by relatives who must cut back drastically on outside work or, in many cases, quit their job. This results in a catch-22 situation: depleting the major source of income that could be applied to skilled memory units.

Although everyone agrees that long-term care is too expensive, don't expect improvement anytime soon. Unfortunately, there is little impetus to come up with solutions that can provide for more people at decreased cost. Why? Let's segue briefly into the *shark tank* again for a grim lesson.

Growth in assisted-living facilities is currently fueled by real estate trusts—*Shark Tank* ventures aimed at generating dependably recession-proof returns for their investors. To achieve these returns, publicity is directed toward wealthy potential purchasers or renters for these upscale units. In typical cases, these units are already spoken for before they are even built. If you are over sixty years of age, you have probably received one or more of the glossy brochures sent out by these upscale assisted-living facilities. The pictures typically feature well-dressed, apparently affluent seniors enjoying a cocktail in a setting one would ordinarily encounter in a high-end condominium.

POSSIBILITIES, NOT DISABILITIES

Do I think the incidence of Alzheimer's will continue to rise and we won't have agents capable of slowing down, if not curing, the disease? I'm confident a beneficial treatment will be available within the next five years if certain obstacles (not all of them scientific) can be overcome.

Slowing the progression of Alzheimer's rather than an outright cure may be the best that can be accomplished within the next five years. Achieving a clinically meaningful improvement would provide immeasurable benefits to patients, along with their caretakers and relatives. What do we mean by "clinically meaningful"? Ah, there's the rub. While doctors and families may hold different opinions, it is worth the effort to reach agreement, according to an expert group convened by the Alzheimer's Association in January 2022.

After more than a year of deliberation, the group concurred that slowing the disease rather than knocking it out altogether might be a more achievable goal for drug trials at the moment. Some are put off by the perception that a clinically meaningful endpoint is more of a socially constructed idea than scientifically determinable. But before pooh-poohing the clinically meaningful endpoint altogether, consider this: slowing the progression of the disease, if achieved early enough, may extend acceptable cognitive function for many years.

But suppose, just for argument's sake, that progress is slower than hoped and we are in the same position in 2030 as we are today. Specifically, what kind of setting would you favor for the housing and maintenance of someone you know, or even perhaps yourself, if afflicted by Alzheimer's?

Let's start with current arrangements. At a certain point— earlier for those who exhibit aggressive behavior or who can no longer dependably control their bowel or bladder—a person with Alzheimer's (whose family can afford it) will be transferred to a specialized facility, usually to a memory unit. He or she will no longer be free to come and go—such units are typically locked. Those who cherish their privacy will be frequently frustrated by procedures they have little ability to influence.

In the better-organized dementia units, due attention will be given to group activity aimed at the laudable goal of increasing socialization. If anyone prefers on a given day not to participate in these activities, that person may be excused—but only after giving reasons that would then need to be accepted. Of course, such a forced arrangement is not entirely bad. The resident may be sick or on the way to developing an illness best diagnosed and treated in the early stages.

I could go on with more details, but I trust you get the point. Although we all recognize that, when it comes to human behavior, one size *doesn't* fit all, you might be persuaded to think otherwise if given the opportunity to observe the daily routines of some Alzheimer and dementia facilities. But does it really have to be that way?

Let's imagine a small community of people with Alzheimer's or one of the other dementias living freely with minimal or even no medication. All are encouraged to plan a goodly portion of their own daily life. They interact each day with medical personnel, but primarily on a need-to basis. Social interaction is encouraged but not forced. Visits from relatives and friends are welcomed rather than grudgingly tolerated out of concern the relatives may interrupt staff procedures.

But in this small community with the average age in the seventies, procedures and protocols, such as they are, remain subservient to the wishes of the residents. Emphasis is placed on the individuals making their own judgments about how they will spend their time and with whom. As during the years before they had to deal with Alzheimer's, they will chose their friends on the basis of reasons they can't always articulate.

If one word best describes this living arrangement, it would be *freedom*—to decide and manage for oneself; to make decisions, plan activities, and mingle with friends of one's own choosing. Essentially, we are talking here about a community that normalizes as much as possible the Alzheimer experience. Does this sound unworkable or, perhaps even worse, an inappropriate and wildly imaginative concept of what Alzheimer's is really all about?

Perhaps it will surprise you to learn that communities similar to what I've described are, in fact, already flourishing in Europe, notably in the Netherlands and France. I'm referring here to the dementia village movement.

The isolated village of Hogeweyk can be found nestled like a Russian doll within the small town of Weesp on the outskirts of Amsterdam. Roughly ten football fields in size, the village is home to about 150 residents with Alzheimer's. In keeping with other small villages in Amsterdam, Hogeweyk is composed of a town square, gardens, theater, and grocery store. Also included are a hair salon, a village green, and a village café and bistro. Residents live in groups of six or seven along with a caretaker within a specially designed home. The homes are designed and furnished much like the traditional Dutch houses the residents originally lived in prior to coming to Hogeweyk.

The caretakers, some 250 geriatric nurses and specialists, remain for the most part in the background. They are trained to recognize incipient problems and intervene appropriately. But for the most part they interfere as little as possible "taking care with both hands behind the back," as one of them described it. For example, the residents freely come and go to the supermarket for short-term needs. But shopping for weekly needs is typically done with a staff person who will work out by discussion what's needed.

The working relationship starts the very day of admission when the incoming person with Alzheimer's or dementia is assured that he or she will not lose the ability to act autonomously; the person will retain a sense of identity and remain capable of individually choosing many of the details of everyday activities. The resident can combat loneliness, boredom, and repetition by freely forming friendships with other residents. The underlying theme of Hogeweyk is self-determination. The communal belief, held by the staff, is that the residents are capable of deciding for themselves. The underlying message isn't "You're sick; your brain is diseased; you're not capable of remembering" but "Let's see how you can make your own decisions about when and what you will eat, when you will go to bed, what you will spend your time doing and with whom." Careful monitoring by the staff helps prevent residents from exceeding the limits imposed by their disease.

Similar villages exist in Canada, Australia, France, and Rome. Whether additional villages will be the wave of the future remains to be seen. The biggest barrier to making self-contained villages the standard in dementia care? Cost, of course. As Josh Planos put it in an *Atlantic* article on dementia villages, "A self-contained village would be extremely difficult to implement in a non-socialized health-care system: meaning that in the United States a facility like Hogeweyk might be impossible for the foreseeable future." (Translation: In the absence of government intervention à la Amsterdam, not much is going to happen until somebody figures out how to make the dementia villages profitable.)

To get more insight into some of the problems involved in opening a facility like Hogeweyk in the United States, I spoke with Jannette Spiering, the founder of Hogeweyk and an adviser to other villages based on the normalizing care-social relationship model.

During her discussions with Americans who wished to speak to her about setting up something similar to Hogeweyk in the United States, she learned firsthand some of the probably insuperable difficulties in setting up an American version of Hogeweyk:

> You Americans are entrapped in a system of your own creation. You have a culture deeply committed to suing organizations if everything doesn't turn out 100 percent as anticipated. This creates an atmosphere of defensiveness about lawsuits, rather than concentrating on innovative ways to help people with dementia. Of course, a resident may slip and fall. In our country, the appropriate medical care is provided free of charge. But in your country, the fall must always be somebody's fault and a lawsuit is filed.

I asked her about the private-investment model. She optimistically suggested that "a sense of social responsibility may induce some investors to accept in the short-term lower returns for their investment."

Unrealistic? Naïve? Probably so. But how nice a world it would be, if Jannette Spiering turns out to be correct.

FINAL THOUGHTS

So what can you do *right now* to decrease your chances of Alzheimer's? It doesn't require putting this book down or even getting up from your chair. But before I tell you, please decide if you agree with this statement: "The older I get, the more useless I feel."

What's your answer? Don't fake it; don't think about it for more than a few seconds.

According to a study of people with MCI carried out at Yale University by researchers Becca Levy and Martin Slade, 35.5 percent of respondents expressed a positive age belief (they disagreed with the statement), while 65 percent espoused a negative age belief (they agreed that they felt useless as they aged). So, if you agreed with the statement, you are in the majority. Unfortunately, this isn't good news for you.

Those with positive age beliefs showed a 30.2 percent greater likelihood of recovery than those with negative age beliefs. This is especially impressive since a return to normal cognition from MCI without an accompanying depression is a very rare occurrence, as I mentioned earlier in the book. As a rule, MCI either progresses to dementia or remains stable.

In a secondary analysis, people with normal cognition rather than MCI were asked "The older I get" question. Those expressing positive beliefs were less likely to develop mild cognitive impairment over the next twelve years compared with those expressing negative beliefs. This held true regardless of age and physical health. The key takeaway is that age beliefs are important and, most importantly, can be modified.

Modifying age beliefs is made possible by the power of self-reflection. Basically, self-reflection includes the active ongoing evaluation of one's thoughts, feelings, and behavior. When psychiatrists and psychologists describe someone as possessing (or lacking) insight, they are referring to self-reflection. And this attribute varies widely within the general population.

Have you ever noticed that some people seem to have no idea what they are experiencing when they get angry or sad? They show all the signs of anger (flushed face, dilated eyes, awkward gestures of impatience), but they don't seem to know that they are angry and will even deny if asked about it. To use a buzz phrase from the 1970s, they are "out of touch with their feelings." A Greek word describes that state: *alexithymia* (*a* means without; *lexi*, words; *thymia*, feelings—lacking words for one's inner experiences, or more correctly, the inability to recognize or identify these experiences).

Those afflicted with this surprisingly common disorder haunt doctor's offices with complaints that there is something physically wrong with their back or their brain or their bowels. They have no insight into their difficulties, especially their own contributions to them. And they do poorly in insight-based psychotherapy. But when the alexithymic person is given antidepressants or antianxiety medications, the physical complaints often melt away.

A person capable of self-reflection, in contrast, recognizes his or her moods and impulses without any need to divert them into physical symptoms (à la the alexithymic).

A study released in 2022 by the Medit-Ageing Research Group included perfectly normal aging adults, as well as aging adults with subjective cognitive decline ("I don't know what's wrong, but my thinking isn't just quite right."). In both groups, high levels of self-reflection were related to increased glucose metabolism and increases in the subject's powers of thinking. Along with low levels of self-reflection, other factors associated with a heightened risk for Alzheimer's and other dementias included depression, anxiety, low conscientiousness, and defensive pessimism (in which ordinary situations are perceived as threatening and unsafe). In another study published in 2020, repetitive negative thinking was added to the list of unhealthy psychological traits leading to cognitive decline.

Neuroscientists have now proven that unhealthy psychological states lead to one or more of the brain changes typical of Alzheimer's. In another study, low conscientiousness, high neuroticism, and repetitive negative thinking were shown to be associated with a greater accumulation of amyloid, the waste product first discovered in the brain of Auguste Deter by neurologist Alois Alzheimer. In contrast, developing one's power of self-reflection leads to increased glucose metabolism in the temporal and parietal lobes, just the opposite to what is seen in Alzheimer's disease. So self-reflection is one psychological trait with scientifically approved effectiveness in warding off Alzheimer's.

I personally find it exciting that studies just within the past year have shown that healthy brain effects can be brought about by positive thinking.

Here is the final lifestyle modification: Don't spend time obsessing about whether you may come down with dementia sometime in the distant future. Instead, follow the current healthy lifestyle guidelines and enjoy your life. Life is to be lived and not constantly fretted about. None of us can exert total control over how our lives will play out or when they will end.

Let's end with general life advice from contemporary philosopher Kieran Setiya. He believes that what's needed to lead a good and satisfying life is the courage to "hope well." "To hope well is to be realistic about probabilities, not to succumb to wishful thinking or be cowed by fear; it is to hold possibilities open."

When it comes to Alzheimer's and the other dementias, let us all hope well.

ACKNOWLEDGMENTS

Contributors and contributions are acknowledged under Sources Consulted.

Special thanks to my office manager–assistant Franziska Bening for keeping everything organized and on schedule.

SOURCES CONSULTED

Abolhasani, Ehsan, Vladimir Hachinski, Nargess Ghazaleh, Mahmoud Reza Azarpazhooh, Naghmeh Mokhber, and Janet Martin. "Air Pollution and Incidence of Dementia: A Systematic Review and Meta-Analysis," *Neurology* 100, no. 2 (January 10, 2023): https://doi.org/10.1212/WNL.0000000000201419.

Addis, Donna Rose, and Lynette J. Tippett. "Memory of Myself: Autobiographical Memory and Identity in Alzheimer's Disease." *Memory* 12, no. 1 (January 2004): 56–74. https://doi.org/10.1080/09658210244000423.

Andrei, Minhai. "Hikikomori, the Japanese Phenomenon of Extreme Social Isolation Is Going Global." ZME Science (website), April 29, 2023. https://www.zmescience.com/feature-post/culture/culture-society/hikikomori-loneliness/.

Ansari, Sam. "The Medical Test Paradox." De Econometrist, January 7, 2021. https://www.deeconometrist.nl/econometrics/the-medical-test-paradox/.

Arlinger, Stig. "Negative Consequences of Uncorrected Hearing Loss—a Review." Supplement, *International Journal of Audiology* 42, no. S2 (August 2023): S17–20. https://doi.org/10.3109/14992020309074639.

Austin, Daryl. "When Looking Back Helps Us Move Forward, or How Nostalgia Can Be Good." *Washington Post*, August 21, 2022. https://www.washingtonpost.com/health/2022/08/21/nostalgia-restorative-first-aid-emotion/.

Beckman Institute. "Reading for Pleasure Strengthens Memory in Older Adults." Neuroscience News, December 6, 2022. https://neurosciencenews.com/reading-aging-memory-22011/.

Bell, Jacob. "Biogen's Alzheimer's Drug Sales Remain Slow as Company Warns of Further Cost Cuts." Biopharma Dive (website), February 3, 2022. https://www.biopharmadive.com/news/biogen -aduhelm-sales-slow-cost-cutting-fourth-quarter/618222/.

Belluck, Pam. "Alzheimer's Drug May Benefit Some Patients, New Data Shows." *New York Times*, November 29, 2022. https://www.nytimes .com/2022/11/29/health/lecanemab-alzheimers-drug.html.

———. "Nuns Offer Clues to Alzheimer's and Aging." *New York Times*, May 7, 2021. https://www.nytimes.com/2001/05/07/us/nuns-offer-clues- to-alzheimer-s-and-aging.html.

Brooks, Megan. "Best Antioxidants to Prevent Age-Related Dementia Identified?" Medscape, May 5, 2022. https://www.medscape.com /viewarticle/973525.

———. "Keto Diet in MS Tied to Less Disability, Better Quality of LIfe." Medscape, March 10, 2022. https://www.medscape.com /viewarticle/970079.

———. "Many Americans Missing an Opportunity to Prevent Dementia." Medscape, May 20, 2022. https://www.medscape.com /viewarticle/974341.

———. "More Evidence Ultraprocessed Foods Detrimental for the Brain." Medscape, August 1, 2022. https://www.medscape.com /viewarticle/978365.

Cassarino, Marcia, Vincent O'Sullivan, Rose Anne Kenny, and Annalisa Setti. "Environment and Cognitive Aging: A Cross-Sectional Study of Place of Residence and Cognitive Performance in the Irish Longitudinal Study on Aging." *Neuropsychology* 30, no. 5 (July 2016): 543–57. https://doi.org/10.1037/neu0000253.

CBS News. "Learning about Alzheimer's from a Study of Nuns." January 31, 2002. https://www.cbsnews.com/news/learning-about-alzheimers -from-a-study-of-nuns/.

Charlesworth, Lara A., Richard J. Allen, Jelena Havelka, and Chris J. A. Moulin. "Who Am I? Autobiographical Retrieval Improves Access to Self-Concepts." *Memory* 24, no. 8 (2016): 1033–41. https://doi.org /10.1080/09658211.2015.1063667.

Chen, Stephanie P., Jay Bhattacharya, and Suzann Pershing. "Association of Vision Loss with Cognition in Older Adults." *JAMA Ophthalmology* 135, no. 9 (September 1, 2017): 963–70. https://doi.org/10.1001/jamaophthalmol.2017.2838.

Clark, John, and Jennifer Williams. "Don't Smoke Pot if You're Old, Experts Warn." My Stateline (website), January 20, 2023. https://www.mystateline.com/news/national/dont-smoke-pot-if-youre-old-experts-warn/.

Cleveland Clinic. "Napping: 3 Proven Health Benefits." September 13, 2021. https://health.clevelandclinic.org/napping-3-proven-health-benefits/.

Cole, Martin G., Lorna Dowson, Nandini Dendukuri, and Eric Belzile. "The Prevalence ond Phenomenology of Auditory Hallucinations among Elderly Subjects Attending an Audiology Clinic." *International Journal of Geriatric Psychiatry* 17, no. 5 (May 2002): 444–52. https://doi.org/10.1002/gps.618.

Cooper, Emily. "Nutrients to Consider with an Autoimmune Disease." IG Living (website), August-September 2022. https://igliving.com/magazine/articles/IGL_2022-08_AR_Nutritients-to-Consider-with-an-Autoimmune-Disease.pdf.

Crawley, Mike. "Why There's Excitement and Skepticism about New Alzheimer's Drug Lecanemab." CBC News, November 26, 2022. https://www.cbc.ca/news/health/alzheimers-disease-drug-lecanemab-human-trial-results-1.6664408.

De Leo, Gianluca, Eleonora Brivio, and Scott W. Sautter. "Supporting Autobiographical Memory in Patients with Alzheimer's Disease Using Smart Phones." *Applied Neuropsychology* 18, no. 1 (January 2011): 69–76. https://doi.org/10.1080/09084282.2011.545730.

Delgado, Carla. "How Naps Improve Memory Performance." *Discover*, July 22, 2022. https://www.discovermagazine.com/health/how-naps-improve-memory-performance.

Dementia Today. "Alzheimer's Timeline Starts 25 Years Before Severe Dementia." Women's Brain Health Initiative, August 8, 2022.

Donovan, Nancy J., and Dan Blazer. "Social Isolation an Loneliness in Older Adults: Review and Commentary of a National Academies

Report." *American Journal of Geriatric Psychiatry* 28, no. 12 (December 2020): 1233–44. https://doi.org/10.1016/j.jagp.2020.08.005.

El Haj, Mohamad, and Pascal Antoine. "Describe Yourself to Improve Your Autobiographical Memory: A Study in Alzheimer's Disease." *Cortex* 88 (January 2017): https://doi.org/10.1016/j.cortex.2017.01.004.

El Haj, Mohamad, Pascal Antoine, and Dimitrios Kapogiannis. "Similarity between Remembering the Past and Imagining the Future in Alzheimer's Disease: Implication of Episodic Memory." *Neuropsychologia* 66 (January 2015): 119–25. https://doi.org/10.1016.j.neuropsycholgia.2014.11.015.

El Haj, Mohamad, Pascal Antoine, Jean Louis Nandrino, and Dimitrios Kapogiannis. "Autobiographical Memory Decline in Alzheimer's Disease, a Theoretical and Clinical Overview. *Ageing Research Reviews* 23, no. B (September 2015): 183–92. https://doi.org/10.1016/j.arr.2015.07.001.

El Haj, Mohamad, Karim Gallouj, and Pascal Antoine. "Autobiographical Recall as a Took to Enhance the Sense of Self in Alzheimer's Disease." *Archives of Gerontology and Geriatrics* 82 (May-June 2019): 28–34. https://doi.org/10.1016/j.archger.2019.01.011.

El Haj, Mohamad, Dimitrios Kapogiannis, and Pascal Antoine. "Phenomenological Reliving and Visual Imagery during Autobiographical Recall in Alzheimer's Disease." *Journal of Alzheimer's Disease* 52, no. 2 (March 16, 2016): 421–31. https://doi.org/10.3233 .JAD-151122.

El Haj, Mohamad, Jean Roche, Karim Gallouj, and Marie-Charlotte Gandolphe. "Autobiographical Memory Compromise in Alzheimer's Disease: A Cognitive and Clinical Overview." *Gériatrie et Psychologie Neuropsychiatrie du Vieillissement* 15, no. 4 (December 1, 2017): 443–51. https://doi.org/10.1684/pnv.2017.0704.

Eustache, Francis, Pascale Piolino, Bénédicte Giffard, Fausto Viader, Vincent de la Sayette, Jean-Claude Baron, and Béatrice Desgranges. "'In the Course of Time': A PET study of the Cerebral Substrates of Autobiographical Amnesia in Alzheimer's Disease." *Brain* 127, no. 7 (July 2004): 1549–60. https://doi.org/10.1093/brain /awh166.

Farr, Evan H. "Robin Williams Would Have Been 70 This Year—What We Now Know about His Lewy Body Dementia." *Elder Law & Estate Planning News* (blog), July 21, 2021. https://www.farrlawfirm.com/dementia/robin-williams-would-have-been-70-this-year-what-we-now-know-about-his-lewy-body-dementia/.

Fisher Center for Alzheimer's Research Foundation. "Vision Loss May Increase Dementia Risk." November 22, 2021. https://www.alzinfo.org/articles/diagnosis/vision-loss-may-increase-dementia-risk/.

Ford, Lucie. "Chris Memsworth to Take a Step Back from Acting after Discovering Alzheimer's Risk." *GQ*, November 21, 2022. https://www.gq-magazine.co.uk/culture/article/chris-hemsworth-alzheimers.

Foster, Russell. "Stop Listening to Sleep Experts." *Wired*, December 22, 2022. https://www.wired.com/story/sleep-health-science/.

Friedman, Richard A. "Ask a Doctor: How Does Marijuana Affect the Adolescent Brain?" *Washington Post*, January 30, 2023. https://www.washingtonpost.com/wellness/2023/01/30/marijuana-adolescent-brain-development/.

George, Judy. "Cognition Boosted by Thinking Positively abouit Aging: Adults with Mild Cognitive Impairment More Likely to Recover if They Held Positive Age Beliefs." MedPage Today (website), April 12, 2023. https://www.medpagetoday.com/neurology/generalneurology/103974.

———. "Dementia Risk Higher for Elite Soccer Players: Is Heading the Ball to Blame?"MedPage Today (website), March 16, 2023. https://www.medpagetoday.com/neurology/dementia/103563.

———. "Dementia Risk May Rise as Air Quality Worsens: Findings Highlight the Importance of Limiting Particulate Matter Pollution." MedPage Today (website), April 5, 2023. https://www.medpagetoday.com/neurology/dementia/103884.

———. "Memory Decline Tied to Lifestyle Factors: Healthy LIfestyle Slowed Memory Loss, Even in APOE4 Carriers." MedPage Today (website), January 25, 2023. https://www.medpagetoday.com/neurology/alzheimersdisease/102811.

———. "NFL Concussion Symptoms Tied to Cognitive Function Decades Later: Players with Concussion Symptoms Performed Worse on Cognitive Tests as They Aged." MedPage Today (website), March 3, 2023. https://www.medpagetoday.com/neurology/headtrauma/103387.

———. "What Happens When Patients Learn about Their Alzheimer's Status?" MedPage Today (website), January 13, 2023. https://www .medpagetoday.com/neurology/alzheimersdisease/102642.

Ghose, Tia. "Robin Williams' Death: What Is Lewy Body Dementia?" Live Science (website), November 3, 2015. https://www.livescience .com/52682-what-is-lewy-body-dementia.html.

Graham, Judith. "A Potential Connection between Dementia and Air Pollution." *Washington Post*, September 19, 2022. https://www.washingtonpost.com /health/2022/09/19/dementia-pollution-connection/.

Gray, Lauren. "Robin Williams' Wife Reveals the Heartbreaking Symptom He Hid from Her." Yahoo!, June 16, 2022. https://www.yahoo .com/now/robin-williams-wife-reveals-heartbreaking-120419320 .html.

Grodstein, Francine, Sue E. Leurgans, Ana W. Capuano, Julie A. Schneider, David A. Bennett. "Trends in Postmortem Neurodegenerative and Cerebrovascular Neuropathologies Over 25 Years." *JAMA Neurol.* 2023; 80(4): 370–376. https://doi:10.1001/jamaneurol.2022.5416.

Gudden, Jip, Alejandro Arias Vasquez, and Mirjam Bloemendaal. "The Effects of Intermittent Fasting and Brain and Cognitive Function." *Nutrients* 13, no. 9 (September 10, 2021): 3166. https://doi.org/10.3390 /nu13093166.

Gunnars, Kris. "The 11 Most Nutrient-Dense Foods on the Planet." Healthline (website), February 23, 2023. https://www.healthline.com /nutrition/11-most-nutrient-dense-foods-on-the-planet#The -bottom-line.

Hailstone, Jamie. "How Air Pollution Can Impact the Mind, Not Just the Lungs." *Forbes*, January 30, 2023. https://www.forbes.com/sites /jamiehailstone/2023/01/30/how-cleaner-air-can-impact-the-mind -not-just-the-lungs/?sh=3d62d3c45d43.

Harvard Health Publishing. "The Benefits of Napping." Healthbeat (blog), May 8, 2012. https://www.health.harvard.edu/healthbeat /the-benefits-of-napping.

———. "The Effects of Marijuana on Your Memory." *Mind & Mood* (blog), November 16, 2021. https://www.health.harvard.edu/mind-and-mood /the-effects-of-marijuana-on-your-memory.

Holland, Kimberly. "What to Expect from Marijuana Withdrawal." Healthline (website), January 17, 2023. https://www.healthline.com /health/marijuana-withdrawal.

Hoosmand, Babak, and Miia Kivipelto. "Antioxidants and Dementia: More Than Meets the Eye." *Neurology* 98, no. 21 (May 24, 2022): 871–72. https://doi.org/10.1212/WNL.0000000000200718.

Janssen, Steve M., David C. Rubin, and Martin A. Conway. "The Reminiscence Bump in the Temporal Distribution of the Best Football Players of All Time: Pelé, Cruijff or Maradona?" *Quarterly Journal of Experimental Psychology* 65, no. 1 (2012): 165–78. https://doi.org /10.1080/17470218.2011.606372.

Kelly, Debra. "10 Unsettling Tales of Sensory Deprivation." Listverse, July 3, 2015. https://listverse.com/2015/07/03/10-unsettling-tales-of -sensory-deprivation/.

Kiper, Dasha. "Dinner with Proust: How Alzheimer's Caregivers Are Pulled into Their Patients' Worlds." *The Guardian*, February 28, 2023. https://www .theguardian.com/society/2023/feb/28/dinner-with-proust-how -alzheimers-caregivers-are-pulled-into-their-patients-worlds.

———. *Travelers to Unimaginable Lands: Stories of Dementia, the Caregiver, and the Human Brain.* New York: Random House, 2023.

Kreimer, Susan. "Hearing Restorative Devices May Have a Beneficial Effect on Cognition." *Neurology Today* 23, no. 2 (January 19, 2023): 1–23. https://doi.org/10.1097/01.nt.0000919240.90683.2c.

Lanese, Nicoletta. "Does the Mediterranean Diet Reduce Dementia Risk? 20-Year Study Hints No." Live Science (website), last modified November 3, 2022. https://www.livescience.com/mediterranean-diet-same -dementia-risk-study.

Linszen, M. M. J., G. A. van Zanten, R. J. Teunisse, R. M. Brouwer, P. Scheltens, and I. E. Sommer. "Auditory Hallucinations in Adults with Hearing Impairment: A Large Prevalence Study." *Psychological Medicine* 49, no. 1 (2019): 132–39. https://doi.org/10.1017/s0033291718000594.

Livingston, Gill, Jonathan Huntley, Andrew Sommerland, David Ames, Clive Ballard, Sube Banerjee, Carol Brayne, Alistair Burns, Jiska Cohen-Mansfield, Claudia Cooper, Sergi G. Costafreda, Amit Dias, Nick Fox, Laura N. Gitlin, Robert Howard, Helen C. Kales, Mika Kivimäki, Eric. B. Larson, Adesola Ogunniyi, Vasiliki Ortega, Karen Ritchie, Kenneth Rockwood, Elizabeth Sampson, Quincy Samus, Lon S. Schneider, Geir, Selbæk, Linda Teri, and Naaheed Mukadam. "Dementia Prevention, Intervention, and Care: 2020 Report of the Lancet Commission." *The Lancet* 396, no. 10248 (August 8, 2020): 413–46. https://doi.org/10.1016/S0140-6736(20)30367-6.

Lomborg, Bjorn. "Climate Change and the Lancet's 'Heat Death' Deception." *Wall Street Journal*, op-ed, November 4, 2022. https://www.wsj.com/articles/the-lancets-heat-death-deception-united-nations-cop-27-cold-study-population-growth-technology-energy-climate-11667580996?mod=article_inline.

Lovato, Nicole, and Leon Lack. "The Effects of Napping on Cognitive Functioning." *Progress in Brain Research* 185 (2010): 155–66. https://doi.org/10.1016/B978-0-444-53702-7.00009-9.

Low, Lee-Fay, and Farah Purwaningrum. "Negative Stereotypes, Fear and Social Distance: A Systematic Review of Depictions of Dementia in Popular Culture in the Context of Stigma." *BioMed Central Geriatrics* 20, no. 1 (November 17, 2020): 477. https://doi.org/10.1186/s12877-020-01754-x.

MacKeen, Dawn. "Worrying if Alzheimer's Will Arrive." *New York Times*, August 16, 2022.

Marcus, Gregory M., David G. Rosenthal, Gregory Nah, Eric Vittinghoff, Christina Fang, Kelsey Ogomori, Sean Joyce, Defne Yilmaz, Vivian Yang, Tara Kessedjian, Emily Wilson, Michelle Yang, Kathleen Chang, Grace Wall, and Jeffrey E. Olgin. "Acute Effects of Coffee Consumption on Health among Ambulatory Adults." *New England Journal of Medicine* 388 (2023): 1092–1100. https://doi.org/10.1056/nejmoa2204737.

Matsushita, Nana, Yuta Nakanishi, Yumi Watanabe, Kaori Kitamura, Keiko Kabasawa, Akemi Takahashi, Toshiko Saito, Ryosaku Komayashi, Ribeka

Tekachi, Rieko Oshiki, Shoichiro Tsugane, Masayuki Iki, Ayako Sasaki, Osamu Yamazaki, Kei Watanabe, and Kazutoshi Nakamura. "Association of Coffee, Green Tea, and Caffeine with the Risk of Dementia in Older Japanese People." *Journal of the American Geriatrics Society* 69, no. 12 (December 2021): 3529–44. https://doi.org/10.1111/jgs.17407.

Mailman School of Public Health. "Calorie Restriction Slows Pace of Aging in Healthy Adults." Columbia University, February 9, 2023. https://www.publichealth.columbia.edu/news/calorie-restriction-slows-pace-aging-healthy-adults.

McGinley, Laurie. "Alzheimer's Drug Sparks Emotional Battle as FDA Nears Deadline on Whether to Approve." *Washington Post*, May 31, 2021. https://www.washingtonpost.com/health/2021/05/31/new-alzheimers-drug/.

McIntosh, Steven. "Chris Hemsworth: Alzheimer's Risk Prompts Actor to Take Acting Break." BBC News, November 21, 2022. https://www.bbc.com/news/entertainment-arts-63668310.

———. "Is It Alzheimer's? Families Want to Know, and Blood Tests May Offer Answers." *Washington Post*, November 17, 2022. https://www.washingtonpost.com/health/2022/11/17/alzheimers-blood-test-research-treatment/?itid=sr_2.

McNeill, Bridgette. "Drinking 2 or More Cups of Coffee Daily May Double Risk of Heart Death in People with Severe Hypertension." American Heart Association, press release, December 21, 2022. https://newsroom.heart.org/news/drinking-2-or-more-cups-of-coffee-daily-may-double-risk-of-heart-death-in-people-with-severe-hypertension.

Mitchell, Richard. "Is Physical Activity in Natural Environments Better for Mental Health Than Physical Activity in Other Environments?" *Social Science & Medicine* 91 (August 2013): 130–34. https://doi.org/10.1016/j.socscimed.2012.04.012.

Mortimer, James A. "The Nun Study: Risk Factors for Pathology and Clinical-Pathologic Correlations." *Current Alzheimer Research* 9, no. 6 (July 2012): 621–27. https://doi.org/10.2174/156720512801322546.

National Academies of Sciences, Engineering, and Medicine. *Social Isolation and Loneliness in Older Adults: Opportunities for the Health Care System*. Washington, DC: National Academies Press, 2020.

National Institute on Aging. "Take Care of Your Senses: The Science behind Sensory Loss and Dementia Risk." National Institutes of Health, January 10, 2023. https://www.nia.nih.gov/news/take-care -your-senses-science-behind-sensory-loss-and-dementia-risk.

———. "What Are Marijuana's Long-Term Effects on the Brain?" National Institutes of Health, April 17, 2023. https://nida.nih.gov /publications/research-reports/marijuana/what-are-marijuanas -long-term-effects-brain.

National Institute on Drug Abuse. "How Does Marijuana Use Affect School, Work, and Social Life?" National Institutes of Health, April 17, 2023. https://nida.nih.gov/publications/research-reports/marijuana /how-does-marijuana-use-affect-school-work-social-life.

Onishi, Norimitsu. "'It's Our Central Park': Uproar Rises over Location of New Toronto Homes." New York Times, February 5, 2023. https:// www.nytimes.com/2023/02/05/world/canada/toronto-greenbelt -development-homes.html.

Osorio, Lolita. "Long-Term Cannabis Use Linked to Dementia Risk Factors." Mescape, April 14, 2022. https://www.medscape.com/viewarticle /972160.

O'Sullivan, Kevin. "Workplace Quality: Is 99.9% Good Enough?" Knowledge Compass (website), April 1, 2018. https://knowledgecompass.com /workplace-quality-is-99-9-good-enough/.

Oudin, Anna, Bertil Forsberg, Annelie Nordin Adolfsson, Nina Lind, Lars Modig, Maria Nordin, Steven Nordin, Rolf Adolfsson, and Lars-Göran Nilsson. "Traffic-Related Air Pollution and Dementia Incidence in Northern Sweden: A Longitudinal Study." *Environmental Health Perspectives* 124, no. 3 (March 2016): 306–12. https://doi.org/10.1289 /ehp.1408322.

Paik, Ji-Sun, Minji Ha, Youn Hea Jung, Gee-Hyun Kim, Kyung-Do Han, Hyun-Seung Kim, Dong Hui Lim, and Kyung-Sun Na. "Low Vision and the Risk of Dementia: A Nationwide Population-Based Cohort Study." *Scientific Reports* 10 (June 2020): https://doi.org/10.1038 /s41598-020-66002-z.

Pang, Linda. "Hallucinations Experienced by Visually Impaired: Charles Bonnet Syndrome." *Optometry and Vision Science* 93, no. 12 (December 2016): 1466–78. https://10.1097/opx.0000000000000959.

Petersen, Ronald C., Paul S. Aisen, J. Scott Andrews, Alireza Atri, Brandy R. Matthews, Dorene M. Rentz, Eric R. Siemers, Christopher J. Weber, and Maria Carrillo. "Expectations and Clinical Meaningfulness of Radomized Controlled Trials." *Alzheimer's & Dementia* (February 7, 2023): https://doi.org/10.1002/alz.12959.

Peterson Colin J. "99.9% Is Good Enough? Sure. Until It Isn't." JIT Outsource, October 15, 2018. https://www.jitoutsource.com/99-9-is -good-enough-sure-until-it-isnt/

Phillips, Matthew C. L. "Fasting as a Therapy in Neurological Disease." *Nutrients* 11, no. 10 (October 2019): 2501. https://doi.org/10.3390/nu11102501.

Rathbone, Clare J., and Chris J. A. Moulin. "Measuring Autobiographical Fluency in the Self-Memory System." *Quarterly Journal of Experimental Psychology* 67, no. 9 (2014): 1661–67. https://doi.org/10.1080/174702 18.2014.913069.

Rauch, Kate. "Alzheimer's Disease, Dementia and the Eye." American Academy of Ophthalmology, July 26, 2022. https://www.aao.org /eye-health/diseases/alzheimers-disease-dementia-eye.

Reed, Nicholas, and Frank Lin. "New Study Links Hearing Loss with Dementia in Older Adults." Johns Hopkins Bloomberg School of Public Health, January 10, 2023. https://publichealth.jhu.edu/2023 /new-study-links-hearing-loss-with-dementia-in-older-adults.

Rees, Matthew. "No First Helpings." Review of *The Oldest Cure in the World: Adventures in the Art and Science of Fasting*, by Steve Hendricks. *Wall Street Journal*, October 6, 2022. https://www.wsj. com/articles/the-oldest-cure-in-the-world-review-no-first-helpings -11665091241.

Robbins, Rebecca. "How Lewy Body Dementia Gripped Robin Williams." *Scientific American*, September 30, 2016. https://www.scientific american.com/article/how-lewy-body-dementia-gripped-robin- williams1/.

Rukovets, Olga. "Social Isolation Is Associated with Future Dementia Risk, New Analysis Finds." *Neurology Today* 22, no. 12 (June 16, 2022): 1–31. https://doi.org/10.1097/01.nt.0000840592.26578.1c.

Sabia, Séverine, Aurore Fayosse, Julien Dumurgier, Alexis Schnitzler, Jean-Philipe Empana, Klaus P. Ebmier, Aline Dugravot, Mika Kivimäki, and Archana Singh-Manoux. "Association of Ideal Cardiovascular Health at Age 50 with Incidence of Dementia: 25 Year Follow-Up of Whitehall II Cohort Study." *British Medical Journal* 366 (2019): 14414. https://doi.org/10.1136/bmj.14414.

Salk Institute. "Deteriorating Neurons Are Source of Human Brain Inflammation in Alzheimer's Disease." News release. December 1, 2022. https://www.salk.edu/news-release/deteriorating-neurons-are-source-of-human-brain-inflammation-in-alzheimers-disease/.

Sauer, Alissa. "What Nuns Are Teaching Us about Alzheimer's." *Our Blog*. Alzheimers.net, n.d. https://www.alzheimers.net/1-09-17-what-nuns-are-teaching-us-about-alzheimers.

Schnaider Beeri, Michal, and Anelyssa D'Abreu. "A Lifelong Perspective for Cognitive Health in Old Age." *Neurology* 99, no. 12 (September 20, 2022): 497–98. https://doi.org/10.1212/wnl.0000000000201069.

Shah, Soleil, Hayden Rooke-Ley, and Erin C. Fuse Brown. "Corporate Investors in Primary Care—Profits, Progress, and Pitfalls." *New England Journal of Medicine* 388 (January 12, 2023): 99–101. https://doi.org/10.1056/nejmp2212841.

Shaw, Gina. "Allergic Disease Linked to Increased Risk of Dementia in Largest Study to Date." *Neurology Today* 22, no. 21 (November 3, 2022): 1–17. https://doi.org/10.1097/01.nt.0000899528.08121.4d.

———. "Dementia and Suicide Risk: Early-Onset Patients, New Diagnoses, and Those with Psychiatric Illness Most at Risk." *Neurology Today* 22, no. 22 (November 17, 2022): https://doi.org/10.1097/01.nt.0000904328.85272.9c.

Shen, Chun, Barbara J. Sahakian, and Jiling Feng. "Author Response: Associations of Social Isolation and Loneliness with Later Dementia." *Neurology* 99, no. 22 (November 2022): 1013. https://doi.org/10.1212/wnl.0000000000201562.

Short, Elizabeth. "Recreational Cannabis Use a Negative for Adolescent Mental Health." Medpage Today (website), May 4, 2023. https://www.medpagetoday.com/pulmonology/smoking/104343.

Snowdon, David A. "Aging and Alzheimer's Disease: Lessons from the Nun Study." *The Gerontologist* 37, no. 2 (April 1997): 150–56. https://doi.org/10.1093/geront/37.2.150.

———. "Healthy Aging and Dementia: Findings from the Nun Study." Pt. 2. *Annals of Internal Medicine* 139, no. 5 (September 2, 2003): 450–54. https://doi.org/10.7326/0003-4819-139-5_part_2-200309021-00014.

Snowdon, David A., L. H. Greiner, and W. R. Markesberry. "Linguistic Ability in Early Life and the Neuropathology of Alzheimer's Disease and Cerebrovascular Disease. Findings from the Nun Study." *Annals of the New York Academy of Sciences* 903 (April 2000): 34–38. https://doi.org/10.1111/j.1749-6632.2000.tb06347.x.

Stine-Morrow, Elizabeth A., Giavanna S. McCall, Ilber Manavbasi, Shukhan Ng, Daniel A. Llano, and Aron K. Barbey. "The Effects of Sustained Literacy Engagement on Cognition and Sentence Processing among Older Adults." *Frontiers in Psychology* 13 (July 11, 2022): https://doi.org/10.3389/fpsyg.2022.923795.

Strikwerda-Brown, Sherie. "Psychological Well-Being: A New Target for Dementia Prevention?" *Neurology* 99, no. 13 (September 27, 2022): https://doi.org/10.1212/wnl.0000000000201110.

Suemoto, Claudia K., Naaheed Mukadam, Sonia M. D. Brucki, Paulo Caramelli, Ricardo Nitrini, Jerson Laks, Gill Livingston, and Cleusa P. Ferri. "Risk Factors for Dementia in Brazil: Differences by Region and Race." *Alzheimer's & Dementia* 19, no. 5 (May 2023): 1849–57. https://doi.org/10.1002/alz.12820.

Swift Yasgur, Batya. "Impaired Vision an Overlooked Dementia Risk Factor." Medscape, April 27, 2022. https://www.medscape.com/viewarticle/972865.

Wainer, David. "Lack of Optimism for Alzheimer's Trials Means There's Little to Lose." *Wall Street Journal*, September 14, 2022. https://www.wsj.com/articles/lack-of-optimism-for-alzheimers-trials-means-theres-little-to-lose-11663121287.

Waldinger, Robert, and Marc Schulz. "The Real Secret of Lifelong Fulfillment." *Wall Street Journal*, January 14–15, 2023. https://www.wsj.com/story/the-real-secret-of-lifelong-fulfillment-6c1d026a.

Walker, Joseph. "New Alzheimer's Drug Shows Positive Results but Side Effects." *Wall Street Journal*, November 29, 2022. https://www.wsj.com/articles/new-alzheimers-drug-shows-positive-results-but-side-effects-11669766449.

Wall Street Journal. "No Country for Alzheimer's Patients." Editorial, March 2, 2023. https://www.wsj.com/articles/biden-administration-alzheimers-treatments-aduhelm-centers-for-medicare-and-medicaid-services-33b6a833.

Weisman, Avery D., and Thomas P. Hackett. "Psychosis after Eye Surgery — Establishment of a Specific Doctor-Patient Relation in the Prevention and Treatment of Black-Patch Delirium." June 26, 1958: *New England Journal of Medicine* 1958; 258: 1284–1289. https://doi.org/10.1056/NEJM195806262582602.

Whalley, Lawrence. "The Cognitive Costs of Social Isolation." *Neurology* 99, no. 2 (June 8, 2022): 47–48. https://doi.org/10.1212/wnl.0000000000200813.

Whitlock Burton, Kelli. "Residential Green Space Linked to Better Cognitive Function." Medscape, May 5, 2022. https://www.medscape.com/viewarticle/973479.

Wnuk, Alexis. "How Does Fasting Affect the Brain?" BrainFacts (website), July 13, 2018. https://www.brainfacts.org/thinking-sensing-and-behaving/diet-and-lifestyle/2018/how-does-fasting-affect-the-brain-071318.

———. "What the Brain of a 104-Year-Old Nun Taught Us about Vascular Dementia." BrainFacts (website), May 14, 2020. https://www.brainfacts.org/diseases-and-disorders/neurodegenerative-disorders/2020/what-the-brain-of-a-104-year-old-nun-taught-us-about-vascular-dementia-051420.

Xie, Echo. "Chinese Brain Researchers Find Evolutionary Clue in Elderly Who Stay Sharp and Have Higher Quality of Life in Old Age." *South China Morning Post*, November 25, 2022. https://www.scmp.com/news/china/science/article/3200943/chinese-brain-researchers-find-evolutionary-clue-elderly-who-stay-sharp-and-have-higher-quality-life.